SNEAKY
uses for
everyday
things

SNEAKY uses for everyday things

Turn a penny into a radio, change milk into plastic, make a dozen STEM projects with everyday things, and other amazing feats

Cy Tymony

Andrews McMeel

PUBLISHING®

Andrews McMeel Publishing
a division of Andrews McMeel Universal
1130 Walnut Street, Kansas City, Missouri 64106

www.andrewsmcmeel.com

20 21 22 23 24 VEP 10 9 8 7 6 5 4 3 2 1

ISBN: 978-1-5248-5330-3

Library of Congress Control Number: 2019954681

Editor: Jean Z. Lucas
Art Director/Designer: Sierra Stanton
Illustrator: Kevin Bremmer
Production Editor: Meg Daniels
Production Manager: Cliff Koehler

ATTENTION: SCHOOLS AND BUSINESSES
Andrews McMeel books are available at quantity discounts with
bulk purchase for educational, business, or sales promotional use. For
information, please e-mail the Andrews McMeel Publishing Special Sales
Department:specialsales@amuniversal.com.

Disclaimer

This book is for the entertainment and edification of its readers. While reasonable care has been exercised with respect to its accuracy, the publisher and the author assume no responsibility for errors or omissions in its content. Nor do we assume liability for any damages resulting from use of the information presented here.

This book contains references to electrical safety that *must* be observed. *Do not use AC power for any projects listed.* Do not place or store magnets near magnetically sensitive media.

Disparities in materials and design methods and the application of components may cause your results to vary from those shown here. The publisher and the author disclaim any liability for injury that may result from the use, proper or improper, of the information contained in this book. We do not guarantee that the information contained herein is complete, safe, or accurate, nor should it be considered a substitute for your good judgment and common sense.

Nothing in this book should be construed or interpreted to infringe on the rights of other persons or to violate criminal statutes. We urge you to obey all laws and respect all rights, including property rights, of others.

Contents

Part I
Sneaky Tricks and Gimmicks

Part II
Sneaky Gadgets and Gizmos

Part III
Security Gadgets and Gizmos

Part IV
Sneaky Survival Techniques

Part V

Sneaky STEM Magnet and Motor Projects

Resources

Acknowledgments

Special thanks go to my agent, Sheree Bykofsky, for her enthusiastic encouragement and for believing in this book from the start. I am also appreciative of the assistance provided by Janet Rosen and Megan Buckley.

I wish to thank Jennifer Fox, my editor at Andrews McMeel, and copy editor, Janet Baker, for their invaluable work.

A warm thank-you goes to Bill Melzer for insights and opinions that helped shape this book.

I am also grateful for the project evaluation assistance provided by Jerry Anderson, Isaac English, Carlos Daza, Sybil Smith, and Serrenity Smith.

And I hope the following is adequate to show my invaluable appreciation and love for Cloise Shaw. Thanks, Mom. I love you.

Introduction

"Life . . . is what we make it."

—William James

_____

You don't have to be 007 to adapt unique gadgets, secure a room from intruders, or get the upper hand over aggressors. Anyone can learn how to become a real-life MacGyver in minutes, using nothing but a few hodgepodge items fate has put at our disposal. Sometimes you have to be sneaky.

Sure, it never hurts to have the smarts of Einstein or the strength of Superman, but they're not necessary with *Sneaky Uses for Everyday Things*. When life puts us in a bind, the best solution is frequently not the obvious one. It'll be the sneaky one.

Solutions to a dilemma can come from the most unlikely sources:

- A motorist stranded with a bad heater-valve gasket made a new one by cutting and shaping the tongue from an old track shoe. It worked well enough to get him home safely.

- U.S. prisoners of war devised a stealthy makeshift radio receiver using nothing more than a razor blade, a pencil, and the wire fence of the prison camp as an antenna.

- Convicts at Wisconsin's Green Bay Correctional Institution scaled the prison walls using rope they braided from thousands of yards of dental floss.

- On September 11, 2001, a window washer trapped in a Twin Towers elevator with five other passengers used his squeegee to pry open the doors and chisel through the wall to escape the inferno.

People rarely think about the common items and devices they use in everyday life. They think even less about adapting them to perform other functions. For lovers of self-reliance and gadgetry, *Sneaky Uses for Everyday Things* is an amazing assortment of more than forty fabulous build-it-yourself projects, security procedures, self-defense and survival strategies, unique gift ideas, and more.

Did you know that the coins in your pocket can generate electricity or receive radio signals? Want to know what household item can identify counterfeit paper currency? How to turn milk into plastic or glue? How to locate directions using the sun or the stars? How to make a compass without a magnet, extract water from thin air, use water to start a fire, or make a ring that can turn on your TV? It's all here. Even wire hangers and coffee creamer-container tops get their moment in the sun.

Sneaky Uses for Everyday Things does not include conventional projects found in most crafts and household hints books. Nor are instructions supplied for first aid, fishing, making a shelter, or spotting edible plants. The Resources section at the back includes lists of books and websites for obtaining science tricks, frugal facts, and camping information.

Sneaky Uses for Everyday Things avoids projects or procedures that require expensive or unusual materials not found in the average home. No special knowledge or tools are needed. Whether you like to conserve resources or like the idea of getting something for nothing, you can use the book as a practical tool, a fantasy escape, or a trivia guide; it's up to you. "Things" will never appear the same again.

The first edition of *Sneaky Uses* was met with great enthusiasm and it has since grown into a nine-title series. Projects from the books and the Sneakyuses.com website have been demonstrated at the annual STEM convention, presented on the National Science Teachers Association (NSTA) website, replicated at hundreds of schools, bookstores, and libraries, as well as featured on local and national TV shows.

Since 2003, we've witnessed a huge rise in interest in the Makers movement and STEM educational initiatives for the do-it-yourself and creative communities. This second edition features all new STEM-oriented projects in magnetism and electricity using everyday things.

Let's get started!

Part I
Sneaky Tricks and Gimmicks

You, too, can do more with less! Many household items you use every day can perform other functions. Using nothing but a few supplies like paper clips, rubber bands, and refrigerator magnets, you can quickly make unique gadgets and gifts.

Want to know how to tell real paper currency from fake? How to make plastic and glue out of milk? Generate electricity from fruits?

If you have an insatiable curiosity for sneaky secrets of everyday things, look no further. The projects that follow can be made in no time. Start your entry into clever resourcefulness here.

The Fear of Small Sums:
Detect Counterfeit Bills

Whether it's a hundred-dollar bill or a one, getting stuck with counterfeit money is a fear many of us have. In the United States in 2018, $70 to $200 million in fake currency was circulated. When counterfeit currency is seized, neither consumers nor companies are compensated for the loss. So what can we do about it? This project describes two methods to tell good currency from bad.

The first method is a careful visual inspection of the bill. Compare a suspect note with a genuine note of the same denomination and series. Look for the following telltale signs:

1. The texture of a legitimate bill should feel rough.
2. Denominations of $5 and higher may have a special security thread and a watermark visible when a light is shining through.
3. Bills $20 and higher may have a color changing denomination mark that appears green and changes to copper when tilted.
4. For additional legitimate bill detection tips, visit the UScurrency.gov website.

The second way to verify paper currency is to test the ink. How can we do this in a sneaky way, at home or in the office? Easy: by using one important feature of the ink used on U.S. currency. A legitimate bill has iron particles in the ink that are attracted to a strong magnet. To verify a bill, obtain a very strong magnet or a rare earth magnet. Rare earth magnets are extremely strong for their small size. They can be obtained from electronic parts stores and scientific supply outlets. See the Resources section at the back of this book.

You can also use small refrigerator magnets, connecting them end to end to create collectively a much stronger single magnet. See Figure 1.

What's Needed
- Dollar bill
- Strong magnet

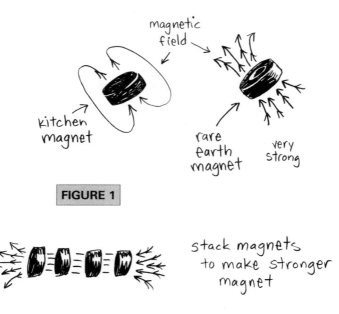

FIGURE 1

stack magnets
to make stronger
magnet

What to Do

Fold the bill in half crosswise and lay it on a table, as shown in Figure 2. Point the strong magnet near the portrait of the president on the bill, but do not touch it. A legitimate bill will move toward the magnet, as shown in Figure 3.

Whenever you doubt the authenticity of paper currency, simply pull out your magnet and perform the magnetic attraction test. If you create the Power Ring shown in Part II of this book, it can be used for currency tests too.

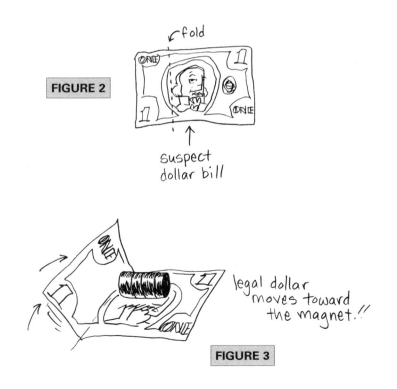

FIGURE 2

fold

suspect
dollar bill

legal dollar
moves toward
the magnet.//

FIGURE 3

Slushy Fun:
Cheap Gel Packs for Swollen Muscles

People in physically demanding jobs and weekend warriors get muscle aches often. When they do, an icy gel pack can relieve the pain and swelling.

Gel packs work to reduce swelling because they can be fitted around joints to cool them thoroughly. You can save money by making your own version—a slushy pack—from everyday things found in the home.

What's Needed
- Water
- Rubbing alcohol
- Watertight freezer bag—typically 6½" x 5⅞"

water

rubbing alcohol

the kind that seals

water proof sandwich bag

FIGURE 1

What to Do

Add 1½ cups water and ½ cup alcohol to the plastic bag and seal it. Ensure that the bag is not overfilled. Place bag in the freezer for 3 hours (see Figure 1).

The fluid inside will not freeze solid. Instead, the alcohol keeps the water flexible and slushy for a better fit (see Figure 2).

When needed, remove the bag from the freezer and apply it to the swollen area as shown in Figure 3. To prevent frostbite or cold burns, place a towel or cloth between the plastic and the skin.

Once you're done, place the slushy pack back in the freezer for future use.

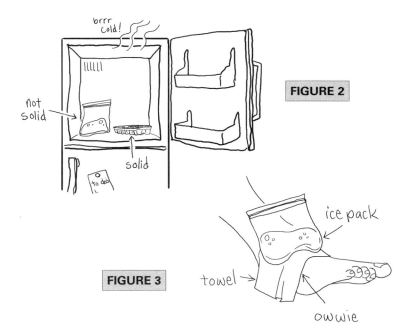

Got Plastic?
Turn Milk into Sneaky Plastic

Have you ever needed plastic molding material for a repair or a craft project? Perhaps you broke off a plastic piece from a toy or an appliance and need to fill in the unsightly gap. Well, you can transform commonplace items in your kitchen into a flexible compound that will do such repairs and even allow you to paint to match.

Believe it or not, you can make a malleable plastic material from plain household milk and only one other ingredient— vinegar. It's easy.

What's Needed
- Milk
- Small pot
- Spoon
- Vinegar
- Strainer
- Jar
- Paper towels

FIGURE 1

What to Do

Pour 8 ounces of milk in a pot and heat it on a stove to 250°F. Let it warm but not boil. Add a tablespoon of vinegar and stir the mixture. Soon, clumps of a solid material will form on the surface. Continue stirring for 3 minutes (see Figure 1).

Place the strainer on top of the jar and pour the mixture through the strainer. Use the spoon to press the clumps and squeeze out the liquid, which is discarded. Remove the material from the strainer and place it on a paper towel. Dab paper towels on top of the material to absorb excess moisture (see Figure 2).

The solid material that has formed is called casein. It separates from milk when an acid, like vinegar, is added. Casein is used in industry to make glue, paint, and some plastics. You can now form the sneaky plastic into shape with a mold or use your hands. Allow the shaped material to dry for 1 to 2 days.

Sneaky plastic has many uses. First, it allows you to recycle spoiled milk that would be discarded anyway. You can make impressions of coins and other small objects. You can shape the plastic into parts to replace items such as broken clips. Or you can make a personalized key ring ornament (see Figure 3).

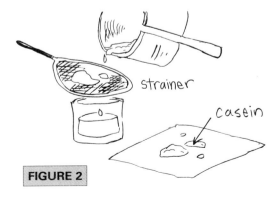

FIGURE 2

Here are some more ideas for putting this plastic compound to use:

Child-proofing items with sharp edges or points
Toy assembly aid (to hold wood and plastic pieces together)
Caulk for small holes in a boat
Pendant holder
Wheels for carts and toys
Tool handle
Material for a spacer or washer
Guitar pick
Bottle cap
Temporary plumbing repairs
Waterproof container
Fishing lure and float
Replacement button

Sneaky plastic can also be used to create a power ring (see Part II under "Superman and Green Lantern Ain't Got Nothin' on Me"). Whether you use the compound for critical repairs or just for fun craft projects, you'll discover it provides plenty of versatility with only a small investment of time.

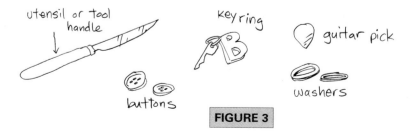

utensil or tool handle

key ring

guitar pick

buttons

washers

FIGURE 3

Need Glue?
Create Sneaky Glue from Milk

If you have an emergency need to secure items together and you're out of glue, don't have a cow. Milk one.

By adding two common ingredients to milk, you can make sneaky glue! When vinegar is added to milk, a sticky substance called casein is formed. By adding baking soda, you can create a gluelike substance.

What's Needed

- Milk
- Small pot
- Spoon
- Vinegar
- Strainer
- Jar
- Paper towels
- Baking soda

FIGURE 1

What to Do

Pour 8 ounces of milk in a pot and heat it on a stove 250°F. Let it warm but not boil. Add a tablespoon of vinegar and stir the mixture. Soon, clumps of a solid material will form on the surface. Continue stirring (see Figure 1).

Place the strainer on top of the jar and pour the mixture through the strainer. Use the spoon to press the clumps and squeeze out the liquid, as shown in Figure 2.

Remove the material from the strainer and place it back in the pot, on the stove. Add ¼ cup of water and a tablespoon of baking soda (see Figure 3). The casein material will begin to bubble. When it stops, use the leftover material as glue.

Note: Wait several hours before using any item secured with sneaky glue.

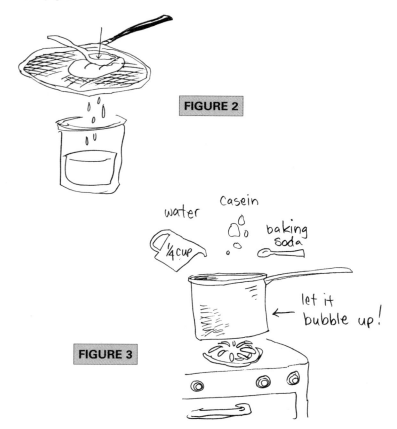

FIGURE 2

water   Casein   baking Soda

¼ cup

← let it bubble up!

FIGURE 3

Getting Wired:
Sneaky Wire Sources Are Everywhere

It will soon become obvious that many projects in this book use electrical wire. In an emergency, you can obtain wire—or items that can be used as wire—from some very unlikely sources.

To test an item's conductivity (its ability to let electricity flow through it), use a flashlight bulb or an LED. (LED is short for light-emitting diode; LEDs are used in most electronic devices and toys as function indicators because they draw very little electrical current, operate with very little heat, and have no filament to burn out.

Lay a small 3-volt watch battery on the item to be tested, as shown in Figure 1. If the bulb LED lights, the item can be used as wire for battery-powered projects.

Ready-to-use wire can be obtained from telephone cord, TV cable, headphone wire, earphone wire, and speaker wire, and also from inside toys, radios, and other electrical devices. (*Note:* Some of these sources have from one to six separate wires inside.)

Wire for projects can also be made from take-out food container handles, twist ties, paper clips, envelope clasps, ballpoint pen springs, fast-food wrappers, and potato chip bag liners.

You can also use aluminum from the following items:

Margarine wrapper
Ketchup or condiment package
Breath mint-container label

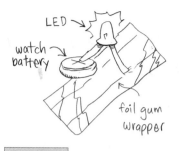

FIGURE 1

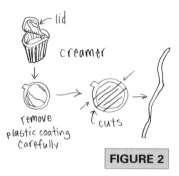

FIGURE 2

Chewing gum wrapper
Trading card packaging
Coffee creamer-container lid

You can cut strips of aluminum material from food wrappers easily enough. With smaller items—such as aluminum obtained from a coffee creamer-container lid—use the sneaky cutting

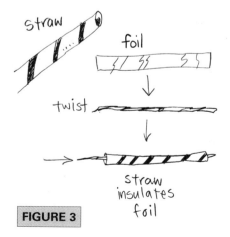

FIGURE 3

pattern shown in Figure 2. Special care must be taken handling the fragile aluminum materials listed. In some instances, aluminum material will be covered by a wax or plastic coating that you may be able to remove.

Note: Wire from aluminum sources is only to be used for low-voltage battery-powered projects.

Figure 3 illustrates how items as small as a cut-up coffee creamer-container lid can be insulated from other items using a straw, hollow stirrer, or paper.

The resourceful use of items to make sneaky wire is not only intriguing, it's fun too.

BONUS APPLICATION:
HOW TO CONNECT THINGS

So far, this project has illustrated how to obtain wires from everyday things. Now you'll learn how to connect the wires to provide consistent performance. A tight connection is crucial in electrical projects; otherwise, faulty and erratic operation may result.

Figure 1 shows a piece of insulated wire. Strip the insulation material away to make a connection to other electrical parts. Remove one to two inches of insulation from the end of the wire; see Figure 2.

To connect the wire to a similarly stripped wire, wrap the stripped ends around each other, as shown in Figure 3's three steps.

When connecting the wire to the end of a stiff lead (like the end of an LED), wrap the wire around the lead and bend the lead back over the wire; see Figure 4.

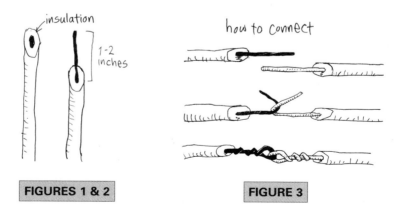

FIGURES 1 & 2

FIGURE 3

To connect wire to the end of a small battery, bend the wire into a circular shape, place it on the battery terminal, and wrap the connection tightly with tape, as shown in Figure 5.

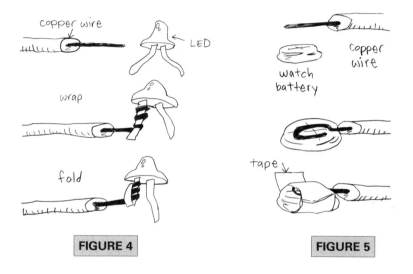

FIGURE 4

FIGURE 5

More Power to You:
Make Batteries from Everyday Things

No one can dispute the usefulness of electricity. But what do you do if you're in a remote area without AC power or batteries? Make sneaky batteries, of course!

In this project, you'll learn how to use fruits, vegetable juices, paper clips, and coins to generate electricity.

What's Needed

- Lemon or other fruit
- Nail
- Heavy copper wire
- Paper clip or twist tie
- Water
- Salt
- Paper towel
- Pennies and nickels
- Plate

What to Do

THE FRUIT BATTERY

Insert a nail or paper clip into a lemon. Then stick a piece of heavy copper wire into the lemon. Make sure that the wire is close to, but does not touch, the nail (see Figure 1). The nail has become the battery's negative electrode and the copper wire is the positive electrode. The lemon juice, which is acidic, acts as

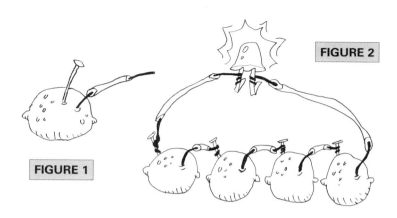

FIGURE 2

FIGURE 1

the electrolyte. You can use other item pairs besides a nail and copper wire, as long as they are made of different metals.

The lemon battery will supply about one-fourth to one-third of a volt of electricity. To use a sneaky battery as the battery to power a small electrical device, like an LED light, you must connect a few of them in series, as shown in Figure 2.

THE COIN BATTERY

With the fruit battery, you stuck the metal into the fruit. You can also make a battery by placing a chemical solution between two coins.

Dissolve 2 tablespoons of salt in a glass of water. This is the electrolyte you will place between two dissimilar metal coins.

Now moisten a piece of paper towel or tissue in the salt water. Put a nickel on a plate and put a small piece of the wet absorbent paper on the nickel. Then place a penny on top of the paper (see Figure 3).

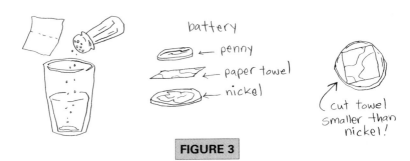

battery
← penny
← paper towel
← nickel

cut towel
smaller than
nickel!

FIGURE 3

In order for the homemade battery to do useful work, you must make a series of them stacked up as seen in Figure 4. Be sure the paper separators do not touch one another.

The more pairs of coins you add, the higher the voltage output will be. One coin pair should produce about one-third of a volt. With six pairs stacked up, you should be able to power a small flashlight bulb, LED, or other device when the regular batteries have failed. See Figure 5. *Power will last up to two hours.*

Once you know how to make sneaky batteries, you'll never again be totally out of power sources.

Six pairs of coin batteries stacked

FIGURE 4

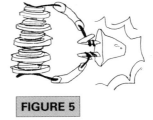

FIGURE 5

You Light Up My Life:
Construct Electronic Greeting Cards

To make an impression on someone, give a gift. For a unique and lasting impression, give a handmade sneaky personalized gift. This project will show you how to use parts from discarded toys and gadgets to make electronic greeting cards, posters, and more.

LEDs are found in most electronic devices and in toys and appliances. They are little lights that indicate whether a device or a function is on. Unlike ordinary lightbulbs, these miniature marvels do not have a filament, produce virtually no heat, consume very little power, and (when properly powered) never burn out!

You can obtain LEDs from old discarded toys and other devices. You will need to cut their leads away from the small circuit board using pliers or wire cutters. You can also purchase LEDs at electronic parts stores locally or from sites shown in the Resources section at the back of the book. You can also obtain blinking LEDs that flash on and off without requiring other components.

What's Needed

- LED
- 3-volt watch battery
- Business card
- Tape
- Wire

business card
folded

insulated wire
(with stripped ends)

tape
rolled up

insulated wire
(with stripped ends)
curled

battery

L E D

L E D

FIGURE 1

What to Do

Since LEDs require 2 to 5 volts to operate, the best compact power supply for them is a 3-volt lithium watch battery. Since AA, C, and D cell batteries provide just 1½ volts each, you would need two cells to provide 3 volts (although larger cells will last longer).

To see how an LED works, press its leads on both sides of a 3-volt battery, as shown in

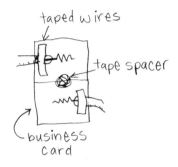

taped wires

tape spacer

business card

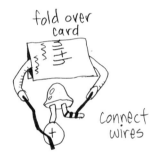

fold over card

connect wires

FIGURE 2

press card

press

you light up my life

Figure 1. If the LED doesn't light, reverse the battery position or reverse the LED leads.

Figure 2 shows how to make a sneaky touch switch with a folded business card, stiff wire, and tape. Tape the wires to the card and roll up a piece of tape to act as a spacer. When properly positioned, a slight press of the folded card will connect the wires and light the LED. See the "Invite the Power" section of Part II for methods of controlling LEDs and other devices with a ring.

LEDs can be mounted on greeting cards and bookmarks, belts and bracelets, behind posters, on trophies—and even on clothing! They can add value to items you might otherwise throw away.

Part II

Sneaky Gadgets and Gizmos

All too often people discard older or broken but still functional high-tech gadgets without realizing what other functions they can serve.

Despite the complexity of radios and other gizmos, this section will illustrate simple, sneaky projects to take advantage of their little-known capabilities.

Want to see how to turn on a TV with your ring or open a room door with a toy car? It's here.

Do you know the sneaky method to make a radio out of a penny or how to turn a screw in an FM radio to magically receive aircraft broadcasts? That's here, too, along with bonus applications.

If you're intrigued with high-tech resourcefulness, the following easy-to-build projects will fascinate and delight.

"Superman and Green Lantern Ain't Got Nothin' on Me": **Make a Power Ring**

What can you do with a commonly worn trinket? More than you might think. Here's a novel project you can make with everyday things that will allow you to activate LEDs, toys—even appliances—with your *ring*!

What's Needed

- Toy ring (with a flat surface)
- Small magnet
- Glue

flat surface

ring

small magnet

What to Do

The power ring works with a magnet attached to its front surface. It is aimed at a magnetically sensitive switch you will make in the next project.

Glue a small magnet to the surface of a toy ring. You can use a refrigerator magnet or, preferably, a small rare earth magnet. A rare earth magnet will allow the power ring to activate devices from a longer distance. Rare earth magnets can be obtained from electronic parts stores and scientific supply outlets (see Figure 1).

You can also create a custom-made ring using the sneaky plastic made from the "Got Plastic?" project in Part I.

Magnet on ring

FIGURE 1

Power Ring Uses

As a Secret Signaling Device. Attach a small toy compass to a door or window with tape. A friend with a power ring can aim the ring from a secret location and the compass needle will spin, signaling that it's your friend.

To Verify Currency. Fold a bill in half and lay it on a table. Point the power ring close to the portrait of the president on the bill. A legitimate bill has iron particles in the ink and will move toward the magnet.

For an Emergency Compass. Need to make a compass? Use your power ring. Obtain a small thin piece of metal (but not aluminum) and stroke the power ring repeatedly in one direction at least fifty times. You can use a straightened paper clip, a needle, a twist tie, or a staple from a magazine. Place the metal on a piece of paper or leaf and let it float on the surface of a

FIGURE 2

container of water. The metal will be magnetized enough to be attracted to the Earth's north and south magnetic poles and act as a sneaky compass.

To Activate Toys and Devices. The next project, "Invite the Power," illustrates how to make a magnetically-activated switch using aluminum foil and a paper clip. With it, only you will be able to activate small lights and buzzers placed inside greeting cards, toys, and gifts as shown in Figure 2.

To Control Appliances. "Invite the Power!" also illustrates how the power ring can be used safely to control household lights and appliances with a device called the X-10® Universal Interface, available from home supply, hardware, and electronic parts stores.

BONUS APPLICATIONS

If you prefer not to activate devices with a ring, a magnet can be mounted in other ways.

Glue a magnet to the end of a plunger dowel (or decorative stick), and you've got a magic wand. Or place a thin rare earth magnet between two business cards and tape them together. Now you can activate selected devices in the style of a magnetic-strip security card; see Figure 3.

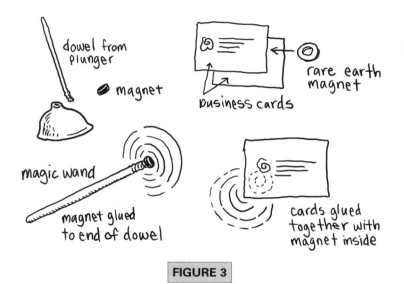

dowel from plunger

magnet

rare earth magnet

business cards

magic wand

magnet glued to end of dowel

cards glued together with magnet inside

FIGURE 3

"Invite the Power!":
Make Power Ring-
Activated Gadgets

What's the good of having a unique power ring without something for it to control? Not much. After you've made a power ring,
as just described, you can use it to activate a variety of devices.
This project will show how to make a magnetically sensitive
switch. With it, the power ring can turn on a variety of devices,
including AC-operated appliances.

What's Needed

- Power ring
- Paper clip or twist tie
- Aluminum foil
- Business card or cardboard
- LED or flashlight bulb
- Wire
- Tape
- Optional X-10® Universal
  Powerflash Interface and
  Appliance Module

power ring

LED

paper twist clip or tie

business card

aluminum foil

insulated wire (with stripped ends)

tape

X-10®
Universal
Interface

screw terminals

button

X-10®
Appliance
Module

What to Do

To create a magnetically activated switch, tape the aluminum foil to the business card. Then tape a piece of wire to the aluminum foil. Roll up a small piece of tape and place it adjacent to the foil, as shown in Figure 1.

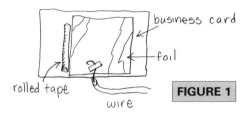

business card

foil

rolled tape

wire

FIGURE 1

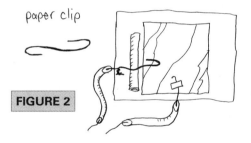

paper clip

FIGURE 2

Next, bend the paper clip into an **S** shape and wrap another

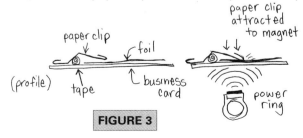

paper clip

foil

(profile)

tape

business card

paper clip attracted to magnet

power ring

FIGURE 3

greeting card

cut-away view

LED

reed switch

battery

Oh Sun Shiny Day

Oh Sun Shiny Day

power ring

FIGURE 4

piece of wire around it. Tape the paper clip near the foil and over the tape so that one end of it rests slightly above, but not touching, the foil; see Figure 2.

Test the switch by aiming the power ring close to the switch. The paper clip should move forward and make contact with the aluminum when the power ring is close; see Figure 3.

Now the ends of the two wires can be attached to a battery and an LED, and when the power ring is aimed at the paper clip the LED will light. Cover the parts with a piece of cardboard and tape the covers closed for protection.

The power ring-activated switch can be used in a variety of applications, such as activation of LEDs in greeting cards, posters. and jewelry; see Figure 4.

Gifts of a Feather
You Make Together:
Build Togetherness Gifts

Here's a novel project, which illustrates that some gifts, like people, need each other. The gift set will illuminate an LED only when they are facing each other.

Togetherness gifts work with the same parts and design used in the power ring project, but each gift has a magnet inside that will activate the other gift's magnetically sensitive sneaky switch.

What's Needed

- Two small strong magnets
- Paper clips
- Aluminum foil
- Cardboard
- Two gift boxes
- Two LEDs
- Wire

small magnet

Cardboard

LED

aluminum foil

paper clip

2 gift boxes

insulated wire (with stripped ends)

What to Do

First, using the instructions given in the "Invite the Power!" section, build two sets of sneaky switches and connect a battery and LED to each as shown in Figure 1.

Mount the sneaky-switch circuits with batteries and LEDs in the two gift boxes. Cut a small hole through which the LEDs can be seen. Then, with the gift boxes facing each other, tape the two magnets so they are positioned opposite the sneaky switches; see Figure 2.

Test the togetherness gift set by placing them close to each other and see if the LEDs will light. If they do not, adjust the positioning of the two magnets until they do.

Last, secure the covers on the gift boxes with tape and write a passage on the outside (or tape a card to each box) with an appropriate message: for instance, *We both need each other.*

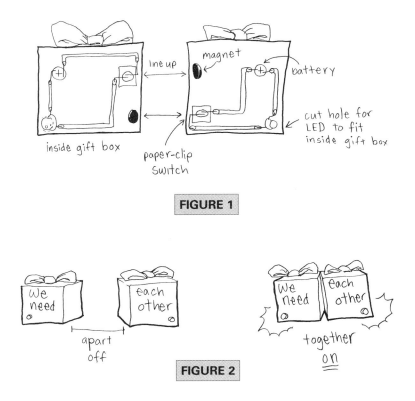

line up

magnet

battery

cut hole for
LED to fit
inside gift box

inside gift box

paper-clip
switch

FIGURE 1

We
need

each
other

apart
off

We
need

each
other

together
on

FIGURE 2

Miniaturizing Mr. Wireless:
Use Him in Remote Places

They say that size matters, but that depends on what your goal is and what you have to work with. If you use the Mr. Wireless device frequently, you'll probably want to miniaturize it for more portable concealed uses. The original package is relatively large to fit the batteries inside, but it can be made smaller.

With a smaller version, Mr. Wireless can be left at various locations for remote monitoring. Imagine the fun and sneaky applications!

What's Needed

- Mr. Microphone toy
- Lithium or other high-capacity watch battery
- Small plastic candy box or mint container
- Tape

Mr. Microphone toy

tape

MINTy Mint

plastic candy box

lithium battery

What to Do

Wireless microphone-type toys vary in design, so the following directions are general in nature.

Remove the screw(s) holding the Mr. Microphone case together. In some instances this will require sliding the two case pieces apart; see Figure 1. Then remove the circuit board, ON/OFF

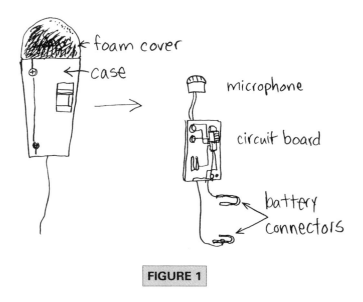

foam cover
case
microphone
circuit board
battery connectors

FIGURE 1

switch, microphone, and battery connectors from the case. Note which connectors are attached to the positive (+) and negative (−) battery terminals.

Assemble your Mr. Wireless parts inside a mint container or other small box, as shown in Figure 2. Tape the lithium watch battery to the connectors and test it with an FM radio to ensure the proper polarity. If it doesn't work, reverse the connectors on the battery. Now your Mr. Wireless is ready for undercover operation.

Keep in mind that miniaturizing the Mr. Wireless device has one disadvantage: the smaller battery used for the modification will have a shorter life.

Remember: The mini Mr. Wireless can be used just about anywhere. But don't forget to bring an FM radio along to monitor its broadcasts.

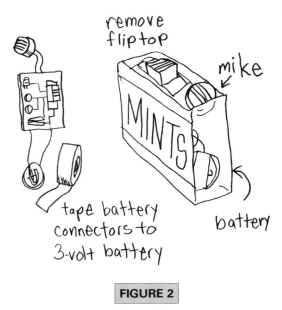

FIGURE 2

Got a Toy Car?
Make a Power Room Door Opener

Who among us hasn't dreamed of having a power door opener as seen in sci-fi and spy movies? This project will show you how to use a small toy car to do the trick. A small wire-controlled car has enough power to push and pull a typical room door back and forth if you know the super-sneaky way to install it.

What's Needed

- Wire-controlled toy car
- Velcro tape, adhesive-backed
- Screwdriver
- Pliers
- Wire hanger

Velcro tape

pliers

screwdriver

wire-controlled toy car

What to Do

This project requires a small wire-controlled toy car, *not* a radio-controlled version. This is to prevent the batteries from running down. (With a radio-controlled car, the remote control and the car's internal receiver have to be in the ON mode, and this drains batteries.)

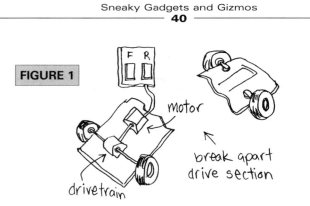

FIGURE 1

motor

break apart
drive section

drivetrain

First, remove the body shell from the toy car with a screw-driver. Then remove the front wheel and axle, as shown in Figure 1. Now, using the Velcro tape, attach the car near the bottom end of the door (see Figure 2).

Using the remote control, see if it can push the door open or closed. If not, reposition the car for more traction. When the proper position is found, you will be able either to move the door with your hand or let the car do it.

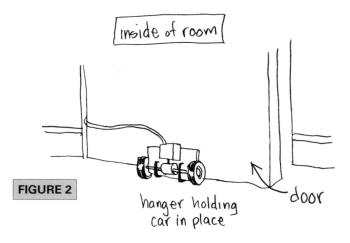

inside of room

FIGURE 2

hanger holding
car in place

door

Optionally, you can break off the entire front part of the chassis so that it takes up less space and cover it with materials for a more appealing look. Mount the remote control outside the door as desired (see Figure 3).

velcroed
to wall

FIGURE 3

Outside of room

BONUS APPLICATION

Substitute a magnetically sensitive sneaky switch for the remote control's switch, and your door can be opened with a power ring. (See the "Invite the Power!" section, earlier in Part II, for details.) Now that's futuristic!

Irrational Public Radio:
Put It Together from Scratch

Ever wonder how a radio works? Would you believe that you can make a sneaky radio with nonelectronic items found in every home? It's true!

Making crystal radios was a popular activity a few generations ago. Even now, many parents show their kids how to make a radio at home that doesn't require AC power or batteries. And there is no danger of getting an electric shock. Although only a few stations will be received, there's nothing like the feeling of tuning in a station for the first time on a radio you put together from scratch.

Normally, building a crystal radio requires electronic components, such as a crystal or germanium diode to act as a detector. This involves going out to purchase a crystal radio kit or hunting for the separate electronic components.

The Irrational Public Radio is made completely out of everyday things. You should be able to receive a strong radio signal in most places in the country. You can use wire from an old telephone cord for the coil, antenna, and ground wires. Instead of a crystal or a diode you will use a penny and a twist tie!

Radio Fun-damentals
Radio stations mix the audio (sound) signal with a carrier signal. The carrier is a high-frequency radio wave that allows for long-range transmission.

Radio receivers have four basic sections: a receiving section, using an antenna and ground wires, to capture radio signals; a tuning section, using a coil of wire, to focus on a specific signal; a detector section, using a diode, to separate the audio signal from the carrier wave; and an audio output section, using a speaker or earphone, to convert the audio signal into sound.

Crystal and diode detectors are electronic devices that conduct electricity in one direction better than the other. Crystals were once used in the detector sections of radio receivers, but diodes are used now. Since crystals and diodes are not considered everyday things, our sneaky radio will use a substitute detector made from a penny and a twist tie.

What's Needed

- Toilet paper tube or small bottle
- Telephone cord, 25 feet or longer (or other thin wire)
- Crystal earphone (not a cell phone headset or in-the-ear earphone)
- Penny
- Paper clips
- Twist tie
- Cardboard (from a shoebox or food container)
- Pliers with insulated handles
- Screwdriver

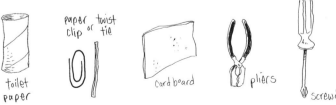

toilet paper

paper clip or twist tie

cardboard

pliers

screwdriver

What to Do

Remove the outer insulation from the telephone wire and separate the wires (see Figure 1). One of the wires will be used for the antenna. Mount it as high as you can in your room. Lay the other end of the wire on the cardboard with a paper clip, as shown in Figure 2. Strip the insulation from the end of the wire and attach it to the paper clip.

Connect another wire, with its end stripped, to a metal water pipe under a sink in your kitchen. This will be the ground wire. Its other end connects to paper clip 2 as shown in Figure 2.

To build the radio tuning coil, remove two inches of insulation from the end of a third wire and put this stripped end through the paper tube. Wind about eighty turns around the paper tube. Secure the wire with tape if necessary. Insert the end of

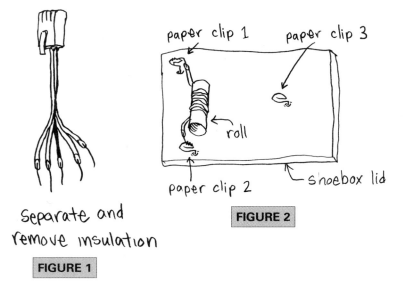

Separate and remove insulation

FIGURE 1

paper clip 1   paper clip 3

roll

paper clip 2   shoebox lid

FIGURE 2

toilet
paper
roll

FIGURE 3

the wire into the other hole on the tube, as shown in Figure 3. Connect both wire ends to paper clips 1 and 2.

Strip an inch of insulation from both wire leads of the earphone. Connect one of the earphone wires to paper clip 2.

Scratch the surface of the penny with the screwdriver and, using pliers, hold it over a flame for three minutes until a black spot appears. After it cools, slip the penny securely under paper clip 3 (see Figure 4).

Strip off the insulation from the twist tie and cut one end into a sharp point. Bend the twist tie and cut one end into a **V** shape and mount the dull end to paper clip 1. The pointed end should press firmly against the burnt surface of the penny, as shown in Figure 4.

Connect the remaining earphone wire to paper clip 3. Your Irrational Public Radio is complete.

Put on the earphone and slowly move the tip of the twist-tie across the surface of the penny until you hear a radio station. Be sure the tip of the twist tie has enough pressure on the coin to maintain contact. If necessary, bend it into position again (see Figure 5).

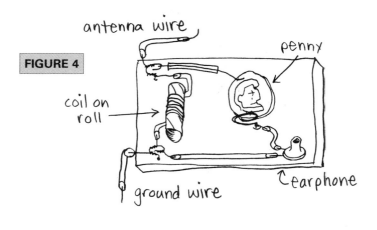

antenna wire

penny

FIGURE 4

coil on roll

ground wire

earphone

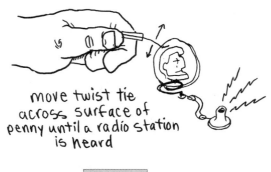

move twist tie across surface of penny until a radio station is heard

FIGURE 5

Con Air:
Convert Your Radio into an Aircraft Broadcast Receiver

"**F**light four-eight-three to LAX tower . . . in need of immediate assistance. We're experiencing an emergency situation."

The VHF aircraft band is filled with fascinating and sometimes critical communications from commercial airliners, the military, and private aircraft. Many aircraft-scanner listeners enjoy the fun and excitement of tuning in tower-to-air conversations that are missed by the general public. Although aircraft radios command premium prices, there is a way to enjoy aircraft broadcasts for free. In fact, if you have an AM/FM radio, you already have an aircraft radio!

Believe it or not, you can easily convert virtually any AM/FM radio into an effective aircraft band receiver. The AM band will remain unchanged. The FM band will have limited reception after the conversion, but it can quickly be returned to its original condition with a small standard screwdriver.

This project should use a small nondigital battery-powered radio. Not too small—ultra-tiny radios can be difficult to open and adjust. But first, some background information on how radios work and what the conversion will do.

Broadcast radio stations transmit a combination of two signals: the audio signal, which includes voice or music, and the carrier signal, which is designed to travel long distances and "carry" the audio signal with it. Voices or music affect (or modulate) the

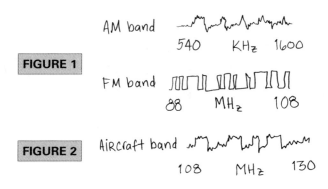

AM band

540   KHz   1600

FIGURE 1

FM band

88   MHz   108

FIGURE 2

Aircraft band

108   MHz   130

audio signal, either by varying its strength, known as *amplitude,* or its *frequency.*

Using voice and music to vary the *amplitude* of the carrier signal is called amplitude modulation—AM. Varying the *frequency* of a carrier signal is the defining feature of *frequency* modulation (FM). See Figure 1 for an illustration of these principles.

An AM radio is designed to tune to a specific range of broadcast stations, discard the carrier signal, and leave the audio signal. Then, just the strength (amplitude) of the signal is forwarded to the radio's speaker or earphone. On the other hand, in an FM radio, just the varying *frequency* of the signal is sent to the radio's speaker or earphone.

Fun Fact 1: AM radios are more sensitive to static because radio interference is a result of the strength of radio waves emanating from lightning or home appliances. FM radios are designed to filter out changes in the strength of the signal, and that's why they produce superior audio quality.

Aircraft radio signals are broadcast on the AM band at a range just above the standard FM band; see Figure 2. What you'll do is convert the FM section of the radio to AM and also allow it to receive signals within the aircraft band. The best thing is that the conversion is simple and does not require any parts!

Fun Fact 2: Before the FCC required digitally-transmitted TV signals in 2009, television signals were both AM and FM; the picture information was broadcast in AM, and the sound signal was broadcast in FM. The two signals were combined and "ride" on the carrier signal. The next time you experience electrical interference in your house, you'll notice that the sound of the TV set is not affected as much as the picture.

What's Needed

- Portable AM/FM radio
- Small flat-blade screwdriver

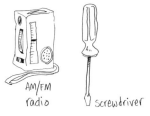

AM/FM radio   screwdriver

What to Do

Test the AM/FM radio to ensure that the battery is strong and that local radio stations can be received loud and clear. Open the back of the radio so you can see the parts on the main circuit board.

Note: Some radio cases snap or slide open while others require you to remove a screw.

Examine the items on the circuit board and look for the part under the tuning dial that selects the stations. This is the variable capacitor. You'll also see two or three small square metal compo-nents. They are filter transformers. Their purpose is to filter out noise and static on the FM band. In essence, they filter out AM signals. Look even closer at the top of the filter transformers and you'll see a screw slot to allow adjustment. Also, one of the filter transformers will have two or three small glass diodes next to it. A diode looks like a clear slender bead with wires on each end.

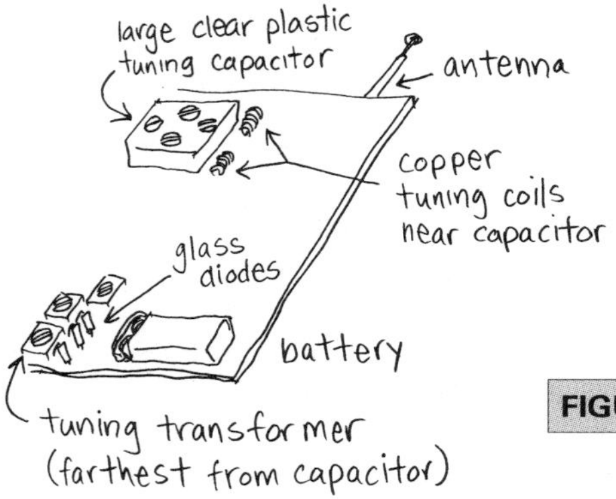

large clear plastic tuning capacitor

antenna

copper tuning coils near capacitor

glass diodes

battery

tuning transformer (farthest from capacitor)

FIGURE 3

This is the filter transformer that we will adjust.

Note: In some radios you may not see the diodes near a tuning transformer, but the adjustment can still be performed. See Figure 3 for an illustration of a typical radio and its component parts.

The first step is to turn the radio on and switch to the FM band. Tune to a spot between stations until you hear a background hiss. Now place your screwdriver in the top of the filter transformer and turn it until the hiss gets as loud as possible. When the hiss is at its highest volume, you have just converted the FM portion of the radio so it can receive AM signals.

One more step is required—you must extend the broadcast range of the FM band. To do this, look at the radio's tuning dial and notice the large square tuning capacitor on the main board. There will be two small copper wire coils next to it. With the screwdriver, spread the small coil windings apart as much as possible without letting them touch another part on the board (see Figure 4). By spreading the coils apart, you have extended

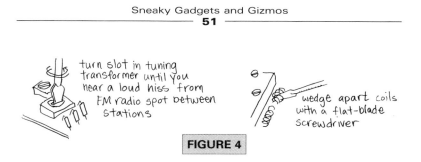

turn slot in tuning transformer until you near a loud hiss from FM radio spot between stations

wedge apart coils with a flat-blade screwdriver

FIGURE 4

the broadcast range from 88 megahertz to 108 MHz and above. Congratulations! Now your radio can receive signals within the aircraft radio band.

Next, tune the dial up and down the FM band. You should notice that the station locations have changed now that the radio is set to receive broadcasts on the aircraft radio band. Take the radio near a local airport and tune across the band. Aircraft messages—from automatic runway signals to tower communiqués—should be audible while tuning the dial.

You can easily convert the radio back to its original state by reversing the two-step process. First, push the small wire coils that are next to the tuning capacitor back to their original positions. Second, dial the radio to a section between stations and retune the filter transformer until the hiss sound is at a minimum.

The next time you visit an airport, take your multiband radio with you and enjoy the fun and intrigue of eavesdropping on aircraft band conversations.

Part III
Security
Gadgets and
Gizmos

Most people like to believe that break-ins and other misfortunes won't happen to them. But ignorance is not always bliss! Nowadays, home, apartment, and hotel security matters are a fundamental concern. This section provides protection devices that can be rigged up to foil assaults on person or property using paper clips, rubber bands, and other household odds and ends.

First, you'll become skilled at using the most innocent-looking items to secure your personal things, like a letter, wallet, or purse. Next, you'll learn how to obtain "see-behind vision" and how to make devices to detect if your doors or windows have been tampered with. Also included is a sneaky project that shows how to rig a disposable camera to identify burglars by taking a thug shot.

Sneaky Ways to Thwart Break-Ins:
Protect Your Fortress from a Man of Steal

Sure, traveling has its charms, but with exotic environments sometimes come mysterious and unexpected dangers. The following projects provide portable and quick-to-set-up security gimmicks for use at home, on a trip, or for an unforeseen stay in a foreign location to thwart or detect window and door break-ins at a moment's notice.

What's Needed

- Rubber bands
- Bubble-wrap material
- String or wire
- Cardboard
- Small bell or chime

rubber bands

bubble-wrap material

Cardboard

small bell or chime

What to Do

To be warned when someone enters a room, place bubble-wrap material under a mat or towel near the entrance. Or place a small bell or chime on the door or window (see Figure 1).

Place an External Sneak Detector (see "Thwart Thieves" later in Part III for details) against a window handle or on a doorsill so it will produce a loud popping sound when activated (see Figure 2).

Optionally, use whatever unbreakable but noisy items are available to place against a door or window to alert you when they are displaced.

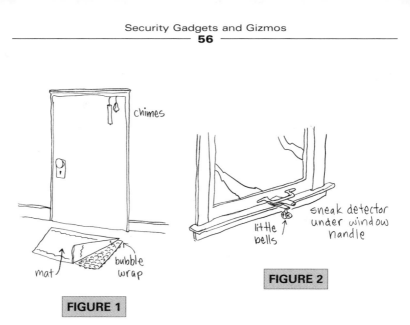

FIGURE 1

FIGURE 2

HOW TO PREVENT BREAKING AND ENTERING

What's Needed

- Wire or strong nylon thread
- Broom, mop handle, plunger dowel, or chair
- Towels
- Tape or rubber bands
- Duct tape

broom or mop handle

rubber bands

tape

wire or nylon thread

What to Do

To protect a door from opening, place a long object, like a broom, mop, or chair, under the doorknob so it's wedged in tight. If necessary, use towels or small pillows to anchor the object to the floor to prevent slippage (see Figure 3).

Even strong nylon thread can prevent a door from opening if it's wrapped tightly from the doorknob to an adjacent window handle or wall light-plate fixture screw. Apply duct tape to all corners of the door to further prevent a break-in (see Figure 4).

Similarly, a window can be secured with an object to prevent it from sliding in its track, as shown in Figure 5.

If a single long object is not available, use shorter ones, like aerosol cans or slender bottles. Position all of the objects end to end and secure them with tape or rubber bands to keep them in line (see Figure 6).

Although a determined burglar can breech these entry inhibitors, they give you valuable time to react and call for help. A chime can be taped to the windows and doors to alert you that a break-in is in progress.

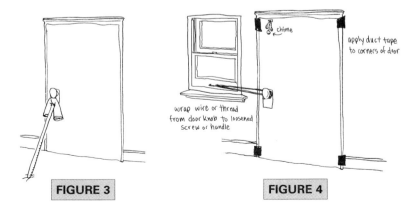

chime

apply duct tape
to corners of door

wrap wire or thread
from door knob to loosened
screw or handle

FIGURE 3

FIGURE 4

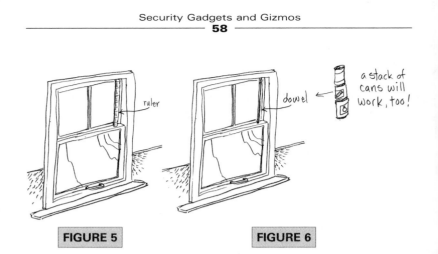

ruler

dowel ←

a stack of
cans will
work, too!

FIGURE 5

FIGURE 6

Foam Alone:
Make a Sneaky Fire Extinguisher

With a little planning and two items found in your kitchen, you can have a lifesaving tool available for emergencies. Having a fire extinguisher on hand gives you peace of mind and can prevent costly damage to your belongings (and possibly save a life). If you ever need a compact and portable fire extinguisher, you can make a sneaky version in just a few minutes using household materials.

What's Needed

- Baking soda
- Vinegar (or lemon juice)
- Jar or plastic soda bottle, 1 pint or larger
- Tissue
- Tape or a rubber band

plastic soda bottle

baking soda

or lemon juice

tape

tissue

What to Do

Fill the bottle halfway with vinegar (or lemon juice). Then form the tissue into a cup shape and poke it into the mouth of the bottle, as shown in Figure 1, holding the sides of the tissue over the rim of the bottle.

Still holding the sides, pour baking soda into the tissue and then place a rubber band or tape around the bottle opening to prevent the tissue from dropping into the bottle; see Figure 2.

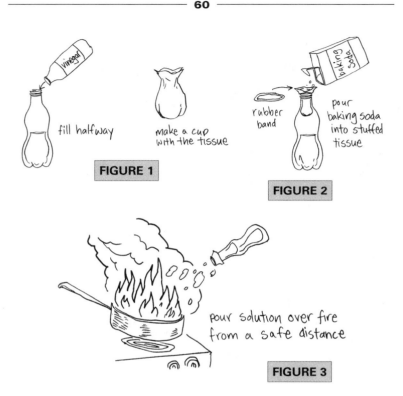

fill halfway

FIGURE 1

make a cup with the tissue

rubber band

pour baking soda into stuffed tissue

FIGURE 2

pour solution over fire from a safe distance

FIGURE 3

Replace the cap on the bottle or place another tissue and rubber band or tape on top of the bottle while it's in storage.

When needed, shake the bottle vigorously and remove the cap. A fire-retardant foam will bubble up to be cautiously poured over a fire. See Figure 3.

Caution: Be particularly careful when in the presence of fire, especially an oil or grease fire.

Gain Sneaky
See-Behind Vision

Forget X-ray vision, heat vision, and microscopic vision. In the real world, whether you're at an ATM or opening your car door, what counts is "see-behind" vision.

Many assaults take place when a person is distracted, and it's impossible to keep looking around while you're fiddling with keys and cards. Luckily, a sneaky remedy is at hand.

What's Needed

- Small mirror or reflective material, about 1½ inches square
- Duct tape
- Large paper clip

What to Do

Bend the large paper clip into the shape shown in Figure 1 and tape it tightly to the small mirror. If the edges of the mirror are

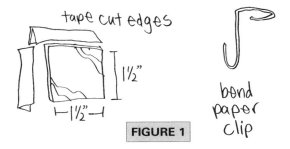

FIGURE 1

sharp, tape them carefully also. Then attach the mirror to a cap or glasses with the other end of the clip, as shown in Figure 2. *Note:* Adjust the position of the mirror by bending the paper clip to improve clarity, if required.

You can use this sneaky vision device three ways: it can be carried in your hand; it can be attached to a cap; or it can be clipped to eyeglasses. Now go ahead and use your see-behind vision for safety and fun.

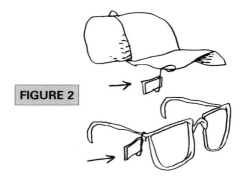

FIGURE 2

Industrious Light Magic:
Make a Sneaky Light in a Pinch

There's an irony in being caught without a light source in a car when there are nearly twenty light bulbs in the average vehicle plus a 12-volt battery under the hood. The only thing required is some wire to connect the battery to the bulb.

Many automotive emergencies happen at night, and most people drive without a working flashlight. You could be caught without ample light to see where a problem is. Light may be needed to find a part on the ground. If you should need to leave the vehicle on a road, a light can be used to signal for help or reveal yourself to motorists so you won't be hit.

This project illustrates two methods of making a small working light from everyday things.

MAKE AN UNDER-HOOD LIGHT

What's Needed

- Small bulb
- Wire

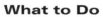

flash-
light
bulb

insulated wire
(with stripped ends)

What to Do

Wire can be obtained from speaker wire connectors or from some other noncritical source in the vehicle. (See "Getting Wired" in Part I for more sources of emergency wire.) As shown in Figure 1, wrap the bare wire around the bulb's side and hold the other end on the battery's positive (+) terminal. Then touch the bulb's bottom to a metal part away from the battery under the hood.

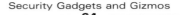

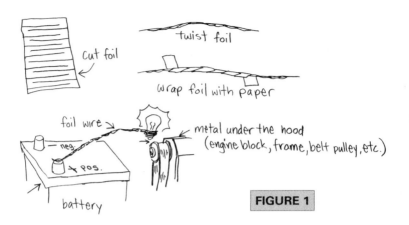

Cut foil

twist foil

Wrap foil with paper

foil wire

neg.

pos.

battery

metal under the hood
(engine block, frame, belt pulley, etc.)

FIGURE 1

If you have two long lengths of wire available, you can attach one end to the bottom of the bulb and the other to the side and have a more flexible light.

MAKE A PORTABLE LIGHT

There's also a way to make a sneaky light for use *away* from the vehicle. When using an automotive bulb, you can use a small 12-volt battery that some portable alarm remote controls use. Otherwise, if you have a flashlight bulb (and the flashlight batteries are depleted) you can use its bulb with batteries you may have on your person.

Note: Do not exceed the bulb's recommended voltage requirement. A, C, D, and AA batteries all supply 1.5 volts. So if a flashlight uses two D-cell batteries, the bulb requires 3 volts to operate. You can then use two C or AA or small watch batteries to light the bulb. Batteries can be found in a car alarm's remote control, a toy, a garage-door opener, or a watch. (In a remote survival situation, batteries from fruits can be used; see "More Power to You" in Part I for details.)

What's Needed
- Battery (that meets bulb voltage requirement)
- Bulb
- Wire

battery

flash-
light
bulb

insulated wire
(with stripped ends)

What to Do
You will need a small length of wire. If no standard wire is available, alternative everyday things that can be used include a paper clip, a twist tie, a metal chain or earring, a ballpoint pen spring, keys, speaker wire, or aluminum foil from a snack bag or coffee creamer-container lid. See Figure 2.

If more than one battery is required, you can wrap paper around them and hold the paper in place with tape, wire, a rubber band, string, or a twist tie.

With the bulb, wire, and batteries available, attach the parts together as shown in Figure 3. If you use small batteries, try to limit your use of the sneaky light to save power.

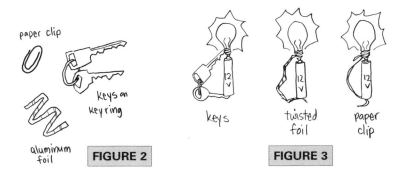

paper clip

keys on
key ring

aluminum
foil

FIGURE 2

keys

twisted
foil

paper
clip

FIGURE 3

Sticky Fingers?
Keep Watch with an Internal Sneak Detector

Are you tired of people prying in your personal items? If you believe that someone is opening your mail or tampering with your belongings, you can plant an Internal Sneak Detector for verification. If you return and the device has snapped apart, you will know someone has opened the item.

What's Needed

- Standard-size envelope
- Paper clip
- Small rubber band
- Piece of strong cardboard
- Scissors
- Pen
- Pliers

paper
clip

rubber
bands

pliers

Cardboard

What to Do

You can use strong cardboard from a tissue box or shoebox. First, cut four small pieces, 3 inches long by 1 inch wide, as shown in Figure 1.

Next, with a pen, punch small holes at the ends of the four cardboard pieces. Then straighten out the paper clip and cut it into four small pieces.

Position the four cardboard pieces and push a piece of paper clip into each hole, as in Figure 2.

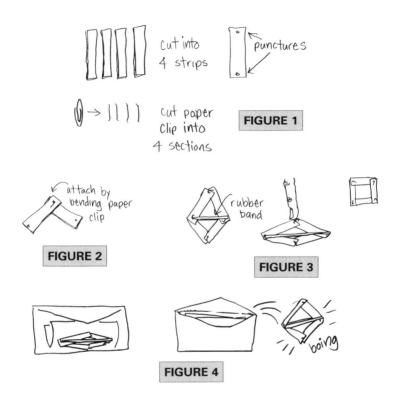

Cut into 4 strips

punctures

Cut paper Clip into 4 sections

FIGURE 1

attach by bending paper clip

FIGURE 2

rubber band

FIGURE 3

FIGURE 4

boing

Attach the rubber band across two of the ends by looping them through the paper clips (see Figure 3).

To use the Internal Sneak Detector, push the other two ends of the rubber band together and place the device into a standard envelope. The rubber band creates spring tension. If the envelope is opened, the detector will spring out, as shown in Figure 4.

You might want to leave a small note, stating that you're aware of the tampering that's been occurring.

Thwart Thieves with the External Sneak Detector

If you use the Internal Sneak Detector, you'll undoubtedly want to try an external version for such larger possessions as a purse, briefcase, book, or laptop computer.

The External Sneak Detector can be placed under an item. It will flip and make a loud snapping sound if the item is moved. You can place a note under it to ward off future mischief.

What's Needed

- Small rubber band
- Piece of strong cardboard
- Scissors
- Glue

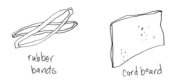

rubber bands

Card board

What to Do

You can use strong cardboard from a tissue box or shoebox. First, from four pieces of cardboard, cut two identical replicas each of shapes A and B, as shown in Figure 1. Glue each pair together for added strength.

Shape B is rectangular. Shape A is the same as B except for a thin notch cut into it from one end to its center; see Figure 1. The exact dimensions of A and B are not critical, but Shape A should extend to half the length of the sneak detector and be wide enough for the small rubber band to fit through easily.

Next, glue pieces A and B together along half of their length, as shown in Figure 2. Glue only the unnotched part of Shape A to

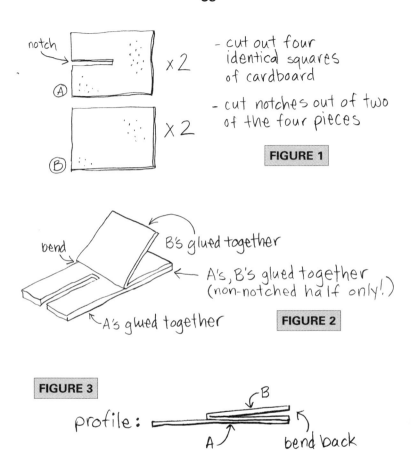

notch

(A) × 2

(B) × 2

- cut out four identical squares of cardboard

- cut notches out of two of the four pieces

FIGURE 1

bend

B's glued together

A's, B's glued together (non-notched half only!)

A's glued together

FIGURE 2

FIGURE 3

profile:

B

A

bend back

Shape B. Once the pieces are glued together, bend the front end of shape B back and forth (see Figure 3).

Now, with pieces A and B flattened out, attach the small rubber band. Be sure to guide the rubber band through the notch. Once it's in place, pull back on the notched half of Shape A until

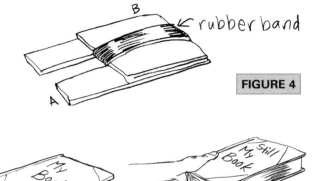

FIGURE 4

FIGURE 5

it bends all the way back. This will stretch the rubber band. If
you let go, it will snap back fast and loud like a mousetrap (see
Figure 4). If this does not happen, use a smaller rubber band or
wrap the band around and through the notch twice.

As seen in Figure 5, the External Sneak Detector, placed
under an object, will fly up and snap with a loud pop when the
object is moved.

Thug Shot:
Capture Break-Ins on Film

A picture can be more valuable than a thousand words . . . in court. If a break-in cannot be thwarted, the next best thing is to record it on film. This project illustrates how to set up a disposable camera so that, if a door or window is breached, you will have a "thug shot" for evidence.

What's Needed

- Disposable camera (non-flash)
- Toy dart gun, with suction-cup tips
- Strong medium-sized rubber band
- Hanger
- Thin strong thread
- Pliers
- Tape
- Cardboard
- Stick-on eyelets

disposable camera

rubber bands

suction cup tip

plastic toy dart gun

hanger

strong thread

tape

pliers

cardboard

stick-on eyelets

What to Do

This project works when a toy gun, triggered by a door or window opening, shoots its dart into a camera's shutter button.

Using pliers, bend the hanger into the shape shown in Figure 1 so that it secures the toy gun and the camera. You should be able to slide the gun in and out of the mount so that a dart, with the suction cup temporarily removed, can be easily reloaded.

Place the toy gun so the dart is aimed a few inches from the camera's shutter button (see Figure 2). Attach one end of the string to the toy gun's trigger and the other end to the door (or window) with stick-on eyelets or suction-cup hooks, as shown in Figure 3.

When you set up the thread and paper clip triggering system, be sure to allow enough room for *you* to slip in and out of the room without setting off the camera. After the correct length and tension of the thread is worked out, when the door (or window) is opened, the thread will pull the toy gun's trigger, causing the dart to shoot the shutter button so the camera takes a photograph.

Since the camera has no flash—a flash would alert the burglar—the room must have sufficient ambient light so the image will be visible on film. Also, since this is a disposable camera, remember to advance the film manually so that it's ready to take a photograph when you leave the room.

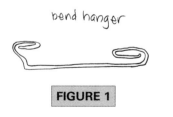

bend hanger

FIGURE 1

← suction cup removed

FIGURE 2

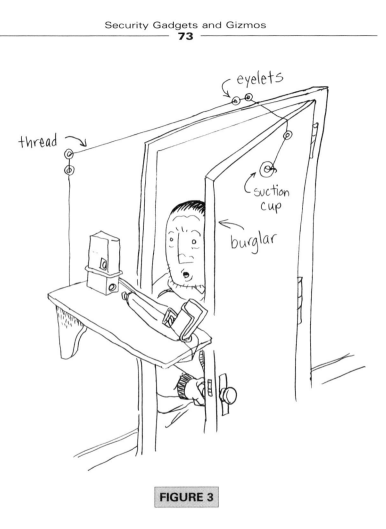

FIGURE 3

Hide and Sneak:
Secure Valuables in Everyday Things

You've seen movies where a character hides something at home and you think, "That's the first place I'd look!" Well, this project will illustrate how to choose sneaky locations that are the *last* places a Man of Steal would look. You don't always have access to a safe deposit box or install alarms on all of your possessions. But you can find sneaky hide-in-plain-sight places to frustrate and waste a thief's time.

Selecting this hiding place generally depends on two factors: the size of the item and the frequency of access required. From a package of soap to a tennis ball, a typical home offers a variety of clever hiding places, as shown in Figure 1. Wrapping your valuables in black plastic bags will further prevent discovery.

With enough time, a tenacious thief can eventually find virtually anything you hide. That's why you should have a room entry alarm installed in combination with sneaky hiding places to reduce the time a thief will spend searching for your valuables.

The Resources section at the back of the book provides a list of security companies marketing clever safes that appear to be soda cans, aerosol sprays, and other everyday things.

pen safe
for emergency
cash

speaker safe

flowerpot
safe

battery compartment
of radio

slit along
seam

tennis
Ball
Safe

store
inside
vacuum
cleaner
bag

FIGURE 1

Part IV
Sneaky
Survival
Techniques

Who hasn't seen *Gilligan's Island* or the movie *Cast Away* and thought, "What would I do if I were lost or stranded somewhere?" Can you imagine being marooned without fire, water, tools, weapons, or a compass? What would you do? How would you survive?

In this section you will find ways to stay afloat in sink-or-swim situations. You will learn how to make a fire, collect water, build a makeshift telescope or magnifying glass, use code-signaling techniques, and make a sneaky emergency light.

You'll see how to survive in the cold and hike in deep snow, and you'll learn direction-finding techniques and how to make a compass.

Part IV concludes with crafty ways to devise makeshift weapons and tools.

Even if you feel that it'll never happen to you, review the text and illustrations. Someday your life may depend on it.

Sneaky Emergency Flotation Devices

If you find yourself in a sink-or-swim scenario, what will you do if a flotation device isn't available? Make a sneaky one from everyday things.

When floating in water, the more you try to keep your head above the surface of the water, the more likely you are to sink. Just lie back and keep your mouth above water.

When you attempt to raise parts of your body above the surface, you lose buoyancy. Luckily, however, you can add to your buoyancy with virtually any empty container that holds air. In some instances, two or more may need to be secured together.

What's Needed

- Plastic bags
- Gas cans
- Large soda bottles
- Other items that will hold air
- String, wire, a belt, or cloth

plastic bags

gas can

large soda bottles

string

What to Do

To make a flotation device of plastic bags, blow the smallest one up, tie a tight knot, and place it in a larger bag (or bags, if available) as shown in Figure 1 to compensate for small holes. Use these inflated bags as water wings to help stay afloat. Figure 2 shows how to tie bags together.

Figure 3 illustrates how to connect two water-holding bottles or jars together. Rest on your back with your head up

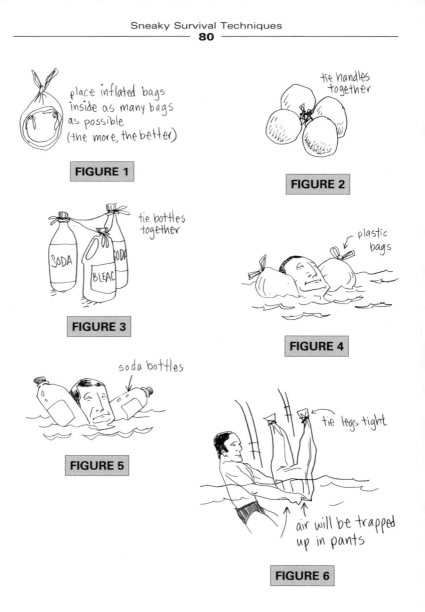

place inflated bags inside as many bags as possible (the more, the better)

FIGURE 1

tie handles together

FIGURE 2

tie bottles together

SODA
SODA
BLEAC

FIGURE 3

plastic bags

FIGURE 4

soda bottles

FIGURE 5

tie legs tight

air will be trapped up in pants

FIGURE 6

(figures 4 and 5).You can also use a log if one is available. Be sure the log will float before laying your body on it (not all wood will float).

There are many other flotation devices that you can devise by using some imagination. Just make sure to test their flotation capabilities before trying to use them.

When no other items are available, your clothing can hold pockets of air to increase your buoyancy. If necessary, remove your pants or shirt, tie the ends of the pant legs or sleeves in knots, and scoop air into them. Hold the other end together firmly with your hands and you should be able to ride above the water with little effort (see Figure 6).

Science Friction:
Six Fire-Making Methods

Experienced campers know how to start a fire without a lighter or matches, but do you? When lost in the wilderness, being able to make a fire can be a lifesaver, both to signal your location and to use for warmth and cooking.

Everyone has heard that it's possible to make fire by rubbing two sticks together. But exactly how do you do this? What if you don't have two dry and properly shaped pieces of wood? Then what do you do?

This project will illustrate six different ways to start a small fire in an emergency. Some of the methods will work and some will not, depending on the resources available, your skill, and your luck. Review all six methods just in case you need to use them one day.

Before you attempt to start a fire, you must have tinder and kindling materials available and understand how to use them. Many people fail to start fires even when they have good matches!

A fire is built in stages. You need first to cause a small fire spark, with one of the methods shown below, to ignite your tinder: small dry items like tissue paper, dead grass, twigs, leaves, lint, or currency. Blow on the tinder carefully, so that it stays lit and grows into a larger fire. Then add kindling—sticks, branches, or thick paper—(*very* carefully, so that you do not suffocate the flame) to keep the fire going. When the kindling is burning, you can add larger logs or other fuel.

METHOD 1:
MAKE A FIRE PLOW

What's Needed

- Hard stick with a blunted tip
- Flat piece of wood
- Tinder
- Kindling
- Knife or sharp-edged rock

stick with blunted end

sharp-edged rock

flat piece of wood

Tinder

What to Do

Using a knife or a sharp rock, scratch a straight indentation in the center of the flat piece of wood about the same width as

FIGURE 1

the blunt stick. Arrange the tinder so air can easily circulate and set it at the foot of the piece of wood, as shown in Figure 1.

Then, in a kneeling position, hold the flat piece of wood between your knees at an angle and move the stick rapidly back and forth in the indentation until friction ignites the fibers of tinder at the base. Mix in more tinder material and fan the smoke until a small fire starts. To keep the fire going, carefully add kindling material.

METHOD 2:
SPARK GENERATION

What's Needed
- Knife or steel
- Sharp-edged rocks
- Tinder
- Kindling

sharp-edged rock

Tinder

What to Do
Use this method with very dry tinder material in a secluded nonwindy environment. Depending on what items are available, strike two rocks together to create a small spark close to tinder material (Figure 2). If a spark catches the tinder, you will see a glow. Carefully blow on it so it turns into a small flame. Fan the material until it starts to smoke and burn (as in Method 1).

If you have an item made of steel, like a knife, scrape it against various rocks until a spark appears.

FIGURE 2

METHOD 3:
MAKE FIRE WITH A BATTERY

You can use the battery from a car or recreational vehicle (or batteries from a flashlight) to start a fire.

What's Needed
- Battery
- Thin wire or metal (from a twist tie or staple, the spring from a ballpoint pen, steel wool, or a flashlight bulb filament)
- Tinder
- Kindling

car battery

Tinder

What to Do
Attach two wires to the battery terminals. With tinder and kindling material at hand, use the other two ends of the wires to create a spark near the tinder (Figure 3). Once the material starts to burn, add kindling material to keep the fire going. If you are using flashlight batteries, put them together, as in Figure 4.

You can also use thin wire strands or other small pieces of metal to start a fire by holding them across the battery (insulate your hands because the wire will get hot) and placing them in the tinder to allow it to burn.

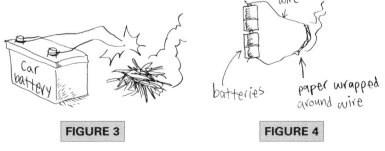

FIGURE 3   **FIGURE 4**

METHOD 4:
MAKE FIRE WITH A LENS

If it's bright and sunny outside, it's possible to use a lens to
focus the heat of the sun on tinder material and start a fire.

What's Needed

- Lens (from eyeglasses—reading
  glasses only—a magnifying glass,
  binoculars, or telescope)
- Tinder
- Kindling

lens

Tinder

What to Do

With plenty of dry tinder available, aim the lens at the tinder until
it starts to smoke. Have other tinder material available to keep
the fire going. When the tinder begins to
burn, add kindling material.
See Figure 5.

FIGURE 5

METHOD 5:
MAKE FIRE WITH A REFLECTOR

What's Needed

- A reflector from a flashlight or car headlight
- Tinder
- Kindling

headlight
reflector

Tinder

What to Do

You can use a light reflector from a flashlight or an old automobile headlight to focus the sun's rays on tinder material. If you use a headlight, carefully break away all the glass.

Position the tinder material in or in front of the reflector for maximum absorption of the sun's rays. With plenty of sunshine available overhead and a little luck, the tinder material will get hot enough to catch fire. See Figure 6.

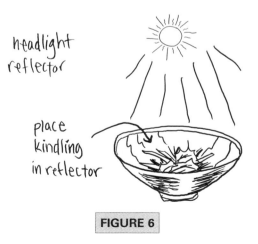

headlight
reflector

place
kindling
in reflector

FIGURE 6

METHOD 6:
MAKE FIRE WITH WATER

When positioned properly, water can act as a lens and focus enough of the sun's heat to ignite tinder.

What's Needed
- Water
- Jar or bottle
- Paper clip or twist tie
- Tinder
- Kindling

paper twist
clip or tie

Tinder

What to Do
Pour about two teaspoons of water into a clear jar or bottle. Tilt the jar so the water rests in a corner at the bottom, then position

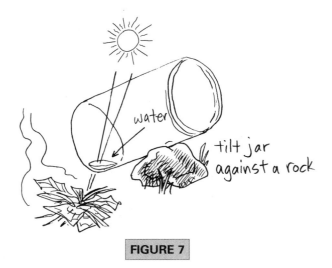

water

tilt jar
against a rock

FIGURE 7

it so the sun's rays shine through the water onto the tinder (see Figure 7) and ignite it.

Note: One of the most intriguing methods is used by tribes in eastern Asia. They hollow out a bamboo cylinder and place small wood shavings at the bottom. When a long wooden rod is forced into the bamboo shaft (like a piston in a car engine), the rapid compression creates enough heat to ignite the wood shavings. The shavings are then poured on tinder to start a flame!

Rain Check:
Two Water-Gathering Techniques

In a survival situation, finding water is crucial; without it, you can only survive a few days. Drinking water from the ocean can be dangerous because of its four-percent salt content. It takes about two quarts of body fluid to rid the body of one quart of seawater. Therefore, by drinking from the ocean, you deplete your body's water, which can lead to death.

Fresh drinking water can be gathered from a variety of sources. This project will show how to gather rainwater and dew from the air.

COLLECTING DEW

What's Needed
- Clean towel or cloth
- Cup, bowl, or other container

What to Do
In the early morning, dew forms on grass, plants, rocks, and other large surfaces near the ground because these items have cooled and water vapor condenses on their surface. The dew can be easily gathered by laying a clean towel on the dew-covered area, dampening it, and wringing the towel out over a bowl. See Figure 1.

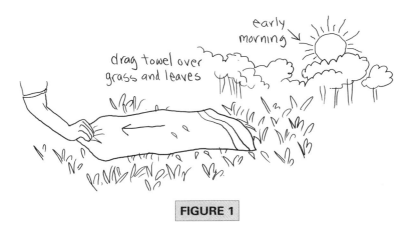

early
morning

drag towel over
grass and leaves

FIGURE 1

GATHERING RAINWATER

Rainwater, when available, is the preferred choice for drinking
because it does not require boiling or purification. It can easily
be collected by setting out items that you may already have.

What's Needed
- Cups, bowls, or other leakproof containers
- Plastic or vinyl material or a nylon jacket
- Rocks

What to Do
Place all available cups and containers where they can fill with
rainwater. If necessary, lay waterproof material—plastic, vinyl
or a waterproof article of clothing—on the ground and secure
the corners with rocks as shown in Figure 2. You can add a few
rocks to the center to create a depression allowing more water

to gather. Or make a container from a large leaf or from coated
paper, as shown in the bonus application "Make a Sneaky Cup."

FIGURE 2

Coming Extractions:
Get Drinking Water from Plants

Water is all around us in the air. The trick in obtaining it is to make it condense on the surface of an object and then collect it in a container.

EVAPORATOR STILL

An evaporator still can be made with a clear or translucent plastic bag and a large plant. It works by allowing the sun to shine through the bag and heat the plant, causing it to give off water vapor through its leaves. The water vapor condenses on the inside surface of the bag and drips down to the bottom. It can then be used for drinking water.

What's Needed
- Large plastic bag, preferably clear
- Plants or leaves

What to Do
Gather green leaves or grass and place them in a plastic bag in a recessed area of the ground, as shown in Figure 1. Select an area where there will be plenty of sunlight. Or choose a plant or leafy tree branch, brush off any excess particles, wrap it in

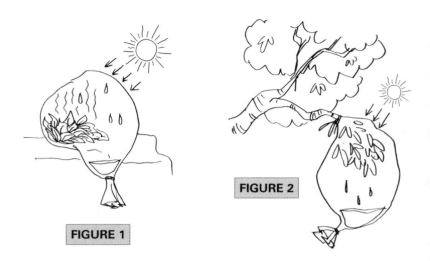

FIGURE 2

FIGURE 1

a plastic bag, and secure its opening with string, wire, or a tight knot; see Figure 2.

As the bag heats up, water from the leaves will evaporate and then condense in the bag as droplets that can be consumed later.

OVERGROUND SOLAR STILL

You can survive up to a month without food but only a few days without water. In the wilderness, there's always a concern about obtaining fresh drinking water. If you are near vegetation and have a large plastic bag available, you can quickly construct a solar still to acquire water.

A solar still uses heat to draw moisture from air, ground, or plants. It then collects the moisture droplets and condenses them into a container for drinking. Solar stills are easy to make, but the amount of water they produce will vary depending on their size, the amount of sunlight, and the terrain.

What's Needed
- Plastic bag, preferably clear, or plastic or vinyl material
- Cup, bowl, or watertight container
- Rocks
- Stick
- Leaves or grass
- Digging utensil

What to Do
The evaporator still proves that water from the air and from plant material can be trapped inside a plastic bag. With an overground solar still, you must dig out an oval or triangular trench and then another around it in an oval shape, as seen in Figure 1. Create the trench on an incline so that water will flow toward the end of the oval section.

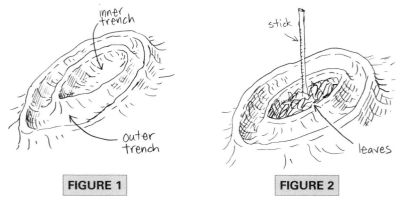

FIGURE 1

FIGURE 2

FIGURE 3

First, place a tall stick in the center of the still and set the leaves or grass inside the center trench; see Figure 2. Next, cover both the stick and both trenches with the plastic bag and hold it in place with rocks. Last, ensure that the bag end is closed and secure. Water from the plants will heat up in the sun, evaporate, condense on the inner surface of the plastic bag, and run down the sides and into the closed end of the bag in the outer oval trench, where it can be poured into a container later; see Figure 3.

BONUS APPLICATION:
H₂ORIGAMI—MAKE A SNEAKY CUP

Gathering or extracting condensed water from the air will be futile without a bowl or cup. You can make a sneaky cup from paper—preferably coated paper from a magazine—or from a large leaf.

What to Do

The following illustrations show a piece of paper with two sides. To clarify the folding technique, one side of the paper is shown as white and the other as gray. Each step is shown in its corresponding illustration figure number.

1. Start with a square piece of paper or large leaf.
2. Fold corner B diagonally on top of corner C.
3. Fold corner A, now point A, down as shown, forming a crease, and then unfold it.
4. Fold corner D, now point D, to the opposite edge, to the place where the first crease hits the edge of the paper.
5. Fold the paper on these two creases.
6. Fold the front (top) flap, corner B, down to cover all the layers. Fold the other flap, corner C, backward (there is only one layer to cover in the back).
7. Open the cup by pulling the front and back apart.

The sneaky cup can be placed underneath a dripping condensation gathering area to save fresh water.

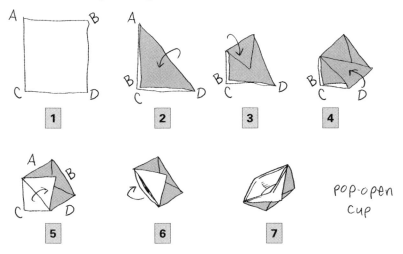

pop-open
cup

Lens Crafter:
Build a Makeshift Telescope

If you're ever lost, it would be convenient, maybe even lifesaving, to be able to see long distances. Didn't bring your binoculars with you? Don't sweat it. Make a sneaky telescope from items you already have. Usually, a telescope requires two lenses. But there is a sneaky way to make a telescope with just one lens and aluminum foil.

What's Needed
- Toilet paper roll or stiff cardboard
- Lens (from a camera, eyeglass, or watch)
- Aluminum foil
- Rubber band
- Paper clip

toilet
paper
tube

lens

rubber
bands

aluminum
foil

paper
clip

What to Do
The aluminum foil from a snack bag or gum wrapper will act as the telescope eyepiece lens. To make this work, you must poke a very tiny round hole in it. To make a good pinhole, stack up several layers of aluminum foil, poke the stack with a pin, separate the layers, and choose the one with the best small round hole. The size of the hole determines whether the images are sharp, blurry, or dim. Test different sizes until you obtain a happy medium.

Wrap the foil, with the pinhole in the center, around one end of the toilet paper roll and secure it with a rubber band, as shown in Figure 1. Cut a slit in the roll at its other end. Attach the paper clip to the lens and slide the lens in the roll by using the paper clip projecting through the slot as a handle; see Figure 2.

You can now use your pinhole telescope to create a zoom-lens effect by moving the lens toward the aluminum pinhole or away. Depending on the distance from pinhole to lens, the scene you see will be either upside down or right side up. It's very complicated to build a zoom-lens telescope with real eyepiece lenses, but if you use a pinhole it becomes simple.

tiny round hole

FIGURE 1

rubber band

slit

FIGURE 2

paperclip

lens

BONUS APPLICATION:
MAKE A SNEAKY MAGNIFYING GLASS

Need to make objects appear larger than they are? When you need to enlarge an image you can easily make a sneaky magnifier. Here are two ways to do it.

What's Needed
- Paper clip, twist tie, or staple
- Clear window envelope or flashlight lens

paper clip or twist tie

window envelope

foil end

lens end

FIGURE 2

What to Do

Method 1. Bend a paper clip into the loop shape shown in Figure 3 and dip the loop in water. As long as it is not too large, the water droplet will stay bonded to the paper clip loop because of surface tension. Look into the water droplet, and the image you see will be magnified.

Method 2. Tear off the window part of the envelope and place a drop of water on it. Now as you look through the water droplet the image will be greatly enlarged; see Figure 4. The farther away you hold the envelope from the object, the larger it will appear.

In situations where you need to see up close, a sneaky magnifier will prove useful.

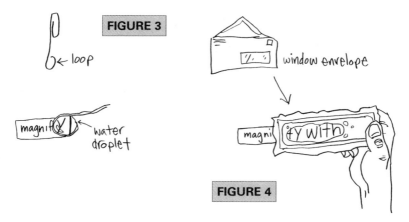

FIGURE 3

← loop

magnif — water droplet

window envelope

magni fy with

FIGURE 4

Smoke and Mirrors:
Sneaky Code Signaling

Being stranded in a remote area can fill you with fear. What's especially frustrating is seeing a plane or vehicle and not being able to get their attention to be rescued.

This chapter will supply various ways to signal for help. With a shiny object that reflects sunlight easily, you can signal to people and vehicles for assistance.

What's Needed

- Reflective materials: canteen, watch, soda can, eyeglasses, mirror, belt buckle, metal pan or cup, or aluminum foil

What to Do

Using a mirror or other shiny object, point the light in one area and away in an SOS pattern (three long flashes, three short, and

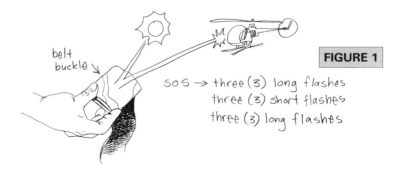

FIGURE 1

SOS → three (3) long flashes
three (3) short flashes
three (3) long flashes

three long). Repeat this sequence of flashes as long as possible while sunlight is available until rescued (see Figure 1).

The *U.S. Army Survival Manual* recommends:

1. Do not flash a signal mirror rapidly, because a pilot may mistake the flashes for an enemy.
2. Do not direct the beam in the aircraft's cockpit for more than a few seconds as it may blind the pilot.

Haze, ground fog, and mirages may make it hard for a pilot to spot signals from a flashing object. If possible, therefore, get to the highest point in your area when signaling. If you can't determine the aircraft's location, flash your signal in the direction of the aircraft noise.

At night you can use a flashlight or a strobe light to send an SOS to an aircraft.

OTHER SNEAKY SIGNALING

When you're lost, use anything and everything as a marker to be seen by aircraft and search parties. Natural materials—snow, sand, rocks, vegetation—and clothing can be used as pointers to spell out distress signals. Follow the Ground-to-Air Emergency Code in laying out your markers.

SYMBOL	MESSAGE
I	Serious Injuries, Need Doctor
II	Need Medical Supplies
V	Require Assistance
F	Need Food and Water
LL	All Is Well
Y	Yes or Affirmative
N	No or Negative
X	Require Medical Assistance
→	Proceeding in This Direction

BODY SIGNAL	MESSAGE
Both arms raised with palms open	"I need help"
Lying on the ground with arms above head	"Urgent medical assistance needed"
Squatting with both arms pointing outward	"Land here"
One arm raised with palm open	"I do not need help"

See Figure 2 for illustrations of these signals.

To show that your signal has been received and understood, an aircraft pilot will rock the aircraft from side to side (in daylight or moonlight) or will make green flashes with the plane's signal lamp (at night). If your signal is received but *not* understood, the aircraft will make a complete circle (in daylight or moonlight) or will make red flashes with its signal lamp (at night).

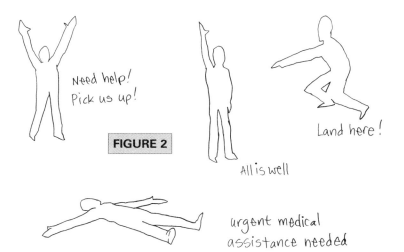

Need help!
Pick us up!

FIGURE 2

Land here!

All is well

urgent medical
assistance needed

Look on the Bright Side:
Make Sneaky Snow Glasses

In daytime, bright snow can blind you and cause damage to the eyes. What's needed is an emergency visor to filter out harmful reflected rays. You can easily make snow glasses out of a variety of things around you.

What's Needed
- Cardboard or leaf
- Material from clothing

What to Do
Cut slits in a piece of cardboard or other material and hold it in front of your eyes. This will cut down on the ultraviolet light reflecting off the snow.

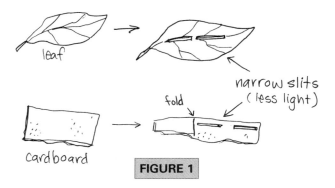

leaf

narrow slits
(less light)

fold

cardboard

FIGURE 1

Sneaky Snowshoes:
Walk on Top of the Snow

Very few of us are so prepared that we have snowshoes readily available in our house or car. In deep snow, your shoes or boots will sink and get wet. You'll waste energy and tire quickly trying to lift your feet out of the snow to take the next steps. Also, you can't see what you are stepping on or how deep you may sink.

What's needed is a way to increase the footprint of the shoe—to spread your weight and allow more snow to provide increased support.

Snowshoes allow you to walk on the surface of the snow. If you're stranded without snowshows in the wilderness, you can quickly make a substitute set out of found items.

What's Needed

- Cardboard, or tree branches and plastic bags
- String, wire, or vine

Card board

plastic bags

string

What to Do

Cut or tear two pieces of heavy cardboard into the shape shown in Figure 1. Cut two holes in the cardboard to feed string, vine, or wire through and wrap it around your shoes and through your laces; see Figure 2.

Making sure that the bulk of the cardboard is in front of the shoe, bend the front upward, as shown in Figure 3. Do not tie

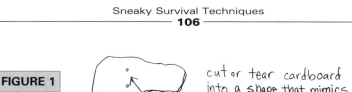

FIGURE 1

holes

cut or tear cardboard into a shape that mimics your shoe, only bigger

FIGURE 2

tie wire across shoe

FIGURE 3

bend front upward

weight focused in one spot

without snow shoe

weight spread out over larger area

with snow shoe

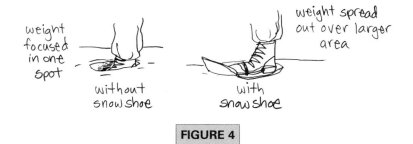

FIGURE 4

the string too tightly. When walking in snow, your foot needs to tilt upward without gathering heavy snow.

If cardboard pieces or similar materials are not available, large tree branches encased in plastic bags can be used in a similar fashion. Now you can walk above the snow, as shown in Figure 4, and keep dry and safe.

Coldfinger:
Where There's a Chill, There's a Way

If you're like most people, you've imagined what it would be like to be stranded in a remote area in cold weather without winter clothes. If you are caught in a wintry climate without proper garments, you could get sick, injured, or worse. This project describes a sneaky way to use the things around you for comfort and safety.

What's Needed
- Leaves or paper
- Plastic bag

What to Do
If it's cold and you do not have the proper outer garments, you will rapidly lose body heat. Heat radiates and is lost from the body, especially from the head, neck, and hands. It is also lost via conduction when you perspire and touch other solid objects. Convection heat loss happens when you are directly exposed to cold winds.

You can reduce heat loss by using whatever is available to cover your head, neck, and hands. Use a plastic bag, newspaper, or undergarment to cover your head, including your mouth. (If the covering is plastic or paper, punch a hole to breathe through.) Covering your mouth as much as possible will direct some of the heat lost through breathing back toward your face as shown in Figure 1.

To create a pocket of air to trap body heat, make a sneaky cold-weather garment with leaves or paper. Assuming you are wearing a long-sleeved shirt and pants, gather leaves from trees or the ground and stuff them in your shirt, as shown in Figure 2. This will create a heat pocket to keep the air warmed by your body. Close sleeve and collar buttons securely.

Similarly, stuff your pants with leaves or paper, draw your socks over your pant legs or knot the ends to stop the leaves from falling out, and keep your hands in your pockets as much as possible; see Figure 2.

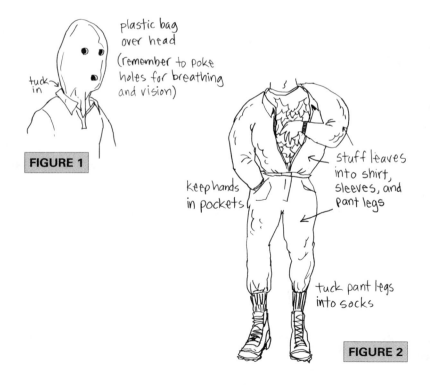

plastic bag over head
(remember to poke holes for breathing and vision)

tuck in

FIGURE 1

keep hands in pockets

stuff leaves into shirt, sleeves, and pant legs

tuck pant legs into socks

FIGURE 2

Lost in Space?
Craft a Compass

If you're ever lost, you'll find a compass is a crucial tool. When markers or trails are nonexistent, a compass can keep you pointed in the right direction to get you back to a line of reference.

A compass indicates Earth's magnetic north and south poles. For a situation where you are stranded without a compass, this project describes three ways of making one with the things around you. For each method, you will need a needle (or twist tie, staple, steel baling wire, or paper clip); a small bowl, cup, or other non-magnetic container; water; and a leaf or blade of grass. How simple is that?

METHOD 1

What's Needed
- Magnet—from a radio or car stereo speaker

What to Do
Take a small straight piece of metal (do not use aluminum or yellow metals), such as a needle, twist tie, staple, or paper clip, and stroke it in one direction with a small magnet. Stroke it at least fifty times, as shown in Figure 1. This will magnetize the needle so it will be attracted to Earth's north and south magnetic poles.

Fill a bowl or cup with water and place a small blade of grass or any small article that floats on the surface of the water. Place the needle on the blade of grass (see Figure 2) and watch

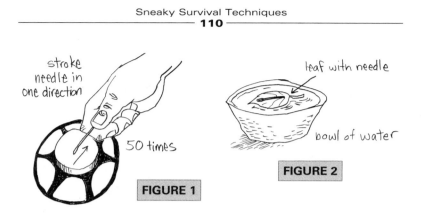

stroke needle in one direction

50 times

FIGURE 1

leaf with needle

bowl of water

FIGURE 2

it eventually turn in one direction. Mark one end of the needle so that magnetic north is determined.

Note: To verify the north direction, see the next section, "Road Scholar: Down-to-Earth Direction Finding."

METHOD 2

What's Needed

piece of silk or other synthetic fabric

- Silk or synthetic fabric— from a tie, scarf, or other garment

What to Do

As in the first method, stroke a needle or paper clip in one direction with the silk material. This will create a static charge in the metal, but it will take many more strokes to magnetize it. Stroke at least three hundred times, as shown in Figure 3. Once floated on a leaf in the bowl, the needle should be magnetized enough to be attracted to Earth's north and south magnetic poles. You may have to remagnetize the sneaky compass needle occasionally.

stroke against silk tie

300 times

FIGURE 3

wire

C battery

paper insulation

needle

FIGURE 4

METHOD 3

What's Needed
• Battery

What to Do
When electricity flows through a wire, it creates a magnetic field. If a small piece of metal, like a staple, is placed in a coil of wire, it will become magnetized.

Wrap a small length of wire around a staple or paper clip and connect its ends to a battery, as shown in Figure 4. (For sneaky battery and wire sources ideas, see "Industrious Light Magic" in Part III.) If the wire is not insulated, wrap the staple with paper or a leaf and then wrap the wire around it.

When you connect the wire to the battery in this manner, you are creating a short circuit—an electrical circuit with no current-draining load on it. This will cause the wire to heat quickly so only connect the wire ends to the battery for short four-second intervals. Perform this procedure fifteen times.

Place the staple on a floating item in a bowl of water, and it will eventually turn in one direction. Mark one end of the staple so that magnetic north is determined.

Road Scholar:
Down-to-Earth Direction Finding

If you're stranded without a magnetic compass, all is not lost. Even without a compass, there are numerous ways to find directions in desolate areas. Two methods are covered here.

METHOD 1:
USE A WATCH

What's Needed
- Standard analog watch
- Clear day where you can see the sun

analog watch

What to Do
The sun always rises in the east and sets in the west. You can use this fact to find north and south with a standard nondigital watch.

If you are in the northern hemisphere (north of the equator), point the hour hand of the watch in the direction of the sun. Midway between the hour hand and 12 o'clock will be south. See Figure 1.

METHOD 2:
USE THE STARS

What's Needed
- A clear evening when stars can be viewed

south

hour hand lines
up with sun

FIGURE 1

What to Do

In the northern hemisphere, locate the Big Dipper constellation in the sky; see Figure 2. Follow the direction of the two stars that make up the front of the dipper to the North Star. (It is about four times the distance between the two stars that make up the

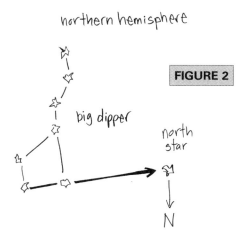

northern hemisphere

FIGURE 2

big dipper

north
star

N

front of the dipper.) Then follow the path of the North Star down to the ground. This direction is north.

In the southern hemisphere, locate the Southern Cross constellation in the sky; see Figure 3. Also notice the two stars below the Cross. Imagine two lines extending at right angles, one from a point midway between the two stars and the other from the Cross, to see where they intersect. Follow this path down to the ground. This direction is due south.

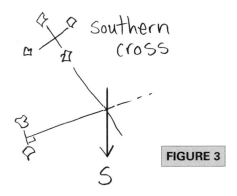

Southern Cross

FIGURE 3

S

Pocket Protectors:
Sneaky Tools
and Survival Kits

How many times have you fumbled in the dark and wished you had a flashlight? Or needed a screwdriver or pliers? Many everyday situations and emergency conditions can be resolved with the proper tools, but most people do not carry a bulky toolbox with them at all times.

This project illustrates a variety of innovative and compact multitools and survival supplies you can carry around in your pocket. For trips away from urban areas, you can assemble a sneaky survival kit that can be carried inside a pen!

MULTITOOLS AND MORE

Compact pocket multitools should be carried, if available. These novel pocket-sized items should become your everyday accessories.

To be prepared, you should always carry these items in your pocket or on your key ring: Figure 1 illustrates the usefulness of this collection.

Multitool (includes four screwdrivers, pliers, cutters, and wire benders)
Mini flashlight
Credit card multitool (includes a compass, magnifying glass, and serrated blade)
Plastic whistle

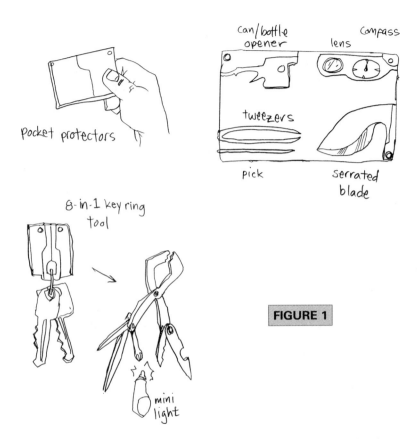

pocket protectors

can/bottle opener

lens

compass

tweezers

pick

serrated blade

8-in-1 key ring tool

mini light

FIGURE 1

For information concerning multitools and gadgets, check the following company websites:

http://www.advanced-intelligence.com
http://www.beprepared.com
http://www.colibri.com
http://www.leatherman.com

http://www.sogknives.com
http://www.spyderco.com
http://www.swissarmy.com
http://www.swisstechtools.com
http://www.topeak.com

SNEAKY SURVIVAL KITS

Why not prepare for the worst when you're traveling? Most people wouldn't want to carry a backpack full of equipment, but if there is a way always to have available the minimal items for a survival situation in a package as small as a mint box or a pen, who will argue against that?

Assuming you already have a multitool, mini flashlight, credit card multitool, and plastic whistle on your key ring, here are a few other items you should take with you when you travel outside of urban areas.

Magnetized needles
Strong nylon thread
Small safety pins (for fishing hooks)
Thin wire
2 small watch batteries, 1½ volts each
Small roll of duct tape
Small roll of aluminum foil (as a signal mirror)
Dental floss
Aspirin
Tiny multivitamins
Tiny candle
5 match heads (placed end to end and wrapped in foil)
$20 bill
Sugar

Believe it or not, you can pack all these items inside a hollowed-out mint container, cigar holder, or fat writing pen! See Figure 1. (When using a pen, remove all of the inner parts and seal both ends.) With these compact tools and supplies, you will greatly improve your odds in survival situations. For more information about assembling survival kits, see the Resources section.

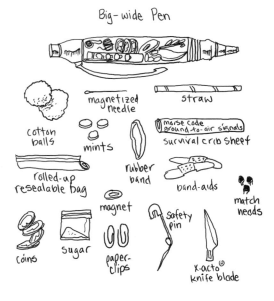

Big-wide Pen

cotton balls

magnetized needle

straw

mints

rolled-up resealable bag

rubber band

morse code ground-to-air signals
survival crib sheet

band-aids

magnet

match heads

coins

sugar

paper-clips

safety pin

X-acto® knife blade

Mint Container

FIGURE 1

Part V
Sneaky STEM Magnet and Motor Projects

STEM stands for Science, Technology, Engineering, and Math. Knowledge in these four academic disciplines is critical for any student wishing to have total control over their own future career. By introducing students to STEM—related concepts in real-life situations, they have a chance to expand their creative thinking and they'll develop a passion for visualizing sneaky re-uses of everyday items all around them.

This section teaches the interplay of magnetism and electricity. The projects that follow provide hands-on experience intended to promote creative thinking, problem-solving, and sneaky resourcefulness. You will learn the science of magnetism and how to construct easy-to-make projects like a compass, an electromagnet, a solenoid . . . and even your own working motor, all using everyday things! Additional design ideas are included to inspire you further to create your own fun motorized inventions.

Sneaky Electromagnetic Fun:
Magnetism Fundamentals

Magnetism and electricity are closely related. A magnet attracts certain ferromagnetic metals. The word "ferrous" means "containing iron" or "like iron," and metals made of iron have the strongest attraction to magnets, along with the metals cobalt and nickel.

The Earth itself is an extremely large magnet, with north and south magnetic poles. (Not to be confused with the north and south geographic poles.) A compass needle points in the direction of Earth's north magnetic pole.

Fun Fact: Since a magnet's poles attract their opposites (for instance, the north pole on one magnet is attracted to another magnet's south pole), the earth's north magnetic pole is really its south geographic pole and its south magnetic pole is really its north geographic pole!

When current flows through a wire, it creates a magnetic field all around the wire. If you put a compass needle near the wire and switch the electricity on or off, you will see the needle move because of the magnetism the wire generates. Electromagnets demonstrate that electricity can create magnetism.

Additionally, you can make ANY metal into an electromagnet by running an electric current through it. If you coil a wire tightly it strengthens the magnetic force—this creates what is called a solenoid. Placing a piece of metal inside the coil will boost the magnetic field even further.

The takeaway of this all is it's the ability of an electric current to turn wires into temporary magnets that makes it possible for us to control motors that can be switched on and off.

SNEAKY WIRE SOURCES

You can round up wire for your projects by reusing old speaker wire, computer network cables, USB cables, AC adapter wire, and video cables.

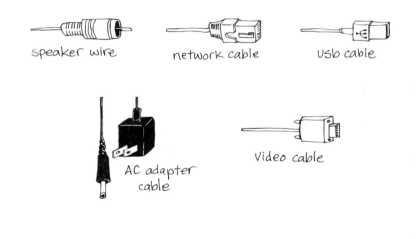

speaker wire   network cable   USB cable

AC adapter
cable

Video cable

SNEAKY MAGNET SOURCES

Find sneaky magnets for your projects from discarded speakers, headphones, earbuds, and refrigerator magnets. You can also obtain strong magnets inside small motors found in discarded toys and electric toothbrushes.

Note: Some of the sneaky motor projects require stronger disc magnets, which can be obtained from craft and hardware stores, home supply outlets, and various online sellers.

earphones

ear buds

speaker

fridge magnet

small toy motor

motor case

Making a Sneaky Compass

A compass uses one of the most useful attributes of magnetism to enable you to locate your orientation and detect nearby magnetic fields.

METHOD 1

What's Needed
- Needle or paper clip
- Cup
- Small piece of Styrofoam or plastic
- Magnet

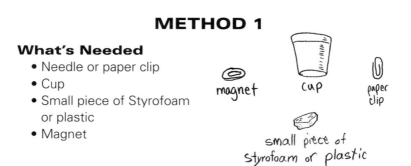

magnet   cup   paper clip

small piece of Styrofoam or plastic

What to Do
To make a simple compass, just stroke a needle (or straightened paper clip) with a strong magnet in the same direction at least fifty times. See Figure 1.

Next, push the needle through the piece of Styrofoam as shown in Figure 2.

Fill a cup with water just below the top. See Figure 3.

Then, place the needle on the water's surface as shown in figure 4. The needle ends will eventually point in the north/south direction. Note which direction is north and mark the Styrofoam piece accordingly.

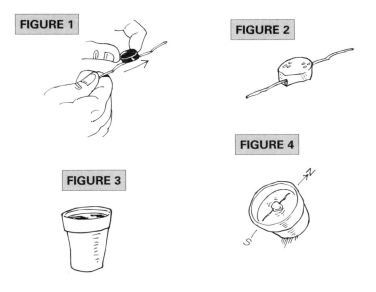

FIGURE 1

FIGURE 2

FIGURE 3

FIGURE 4

METHOD 2

What's Needed
- Strong disk magnet
- Plastic spoon
- Tape

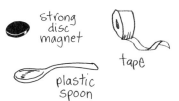

strong
disc
magnet

plastic
spoon

tape

What to Do
Break the handle off the plastic spoon as shown in Figure 1.

Apply tape to the sides of a strong disk magnet. See Figure 2.

While holding the magnet upright, tape it to the spoon with the ends aligned with the short axis of the spoon as shown in Figure 3.

Spin the spoon on a smooth surface (see Figure 4) and you'll find that soon one side will stop, eventually pointing north as shown in Figure 5.

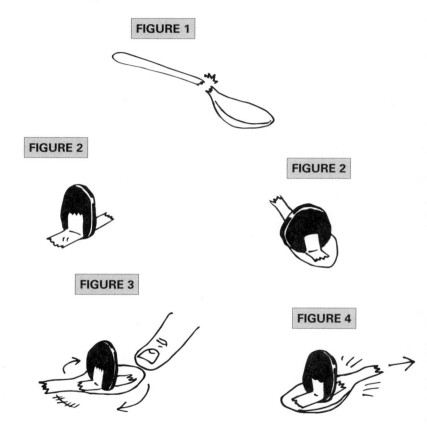

FIGURE 1

FIGURE 2

FIGURE 2

FIGURE 3

FIGURE 4

Electromagnetism in Motion

What's Needed

- 4-cell battery holder
- Batteries
- Compass
- 3 feet or more of insulated copper wire
- Tape
- Pen

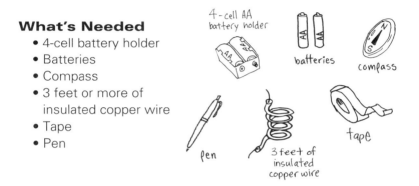

4-cell AA battery holder

batteries

compass

pen

3 feet of insulated copper wire

tape

What to Do

First, wind the copper wire around the pen, leaving five inches of loose wire on each end. Strip the insulation off both ends of the wire so the ends are exposed, as shown in Figure 1.

Wrap tape securely around the wire coil and then slip it off the pen. See Figure 2.

Now tape one end of the wire coil to the bottom or negative (-) battery holder terminal and then place the wire coil near the compass. Affix tape over the other end of the coil's wire, then over the battery holder's positive (+) terminal, so your finger does not touch the bare wire. Press the tape and you will see the compass needle move. See Figure 3.

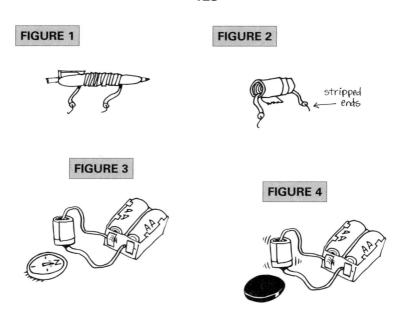

FIGURE 1

FIGURE 2

stripped
ends

FIGURE 3

FIGURE 4

Move the compass away and place the coil of wire on top of a strong magnet. When you connect the battery wire, you should see the coil jump. If it does not, turn the magnet over and test it again. See Figure 4.

Note: Only connect the wire coil to the battery for a few seconds at a time. If left connected for too long, it may get very warm.

Making a Sneaky Solenoid

What's Needed

- 4-cell battery holder
- Batteries
- Paper clip
- 3 feet or more of insulated copper wire
- Tape
- Pen

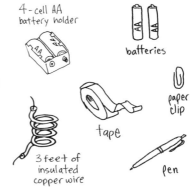

4-cell AA battery holder

AA

batteries

paper clip

tape

3 feet of insulated copper wire

pen

What to Do

Wrap fifty turns of insulated wire around the pen, secure it with tape, and slide the coil off the pen. See Figure 1.

Strip both ends of the wire as shown in Figure 2.

Tape one end of the wire coil to the bottom or negative (-) battery holder terminal as shown in Figure 3.

Affix tape over the other end of the coil's wire and then over the battery holder's positive (+) terminal so your finger does not touch the bare wire. Now place a straightened paper clip halfway into the coil. See Figure 4.

When you press the loose end of wire on the battery holder's positive (+) terminal, the paper clip will slide into the center of the coil. If it does not, push it a little farther into the coil and try again as seen in Figure 5.

Why does this happen? The wire coil creates a strong magnetic field and attracts the paper clip. This is how magnetic door locks work.

Note: Do not keep this connected for more than a few seconds at a time.

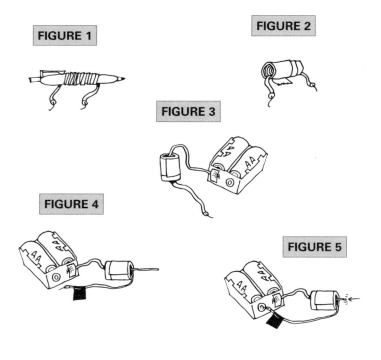

FIGURE 1

FIGURE 2

FIGURE 3

FIGURE 4

FIGURE 5

How to Make an Electromagnet

What's Needed

- AA, C, or D battery
- Several paper clips
- 1 long steel screw or nail
- 3 feet or more of insulated copper wire
- Tape
- Pen

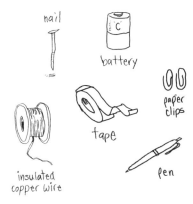

nail

battery

paper clips

tape

pen

insulated copper wire

What to Do

Wrap five feet of wire around the pen, leave six inches free on each end and secure it with tape. See Figure 1.

Next, tape one end of the wire coil to the bottom or negative (-) battery terminal. Then affix tape over the other end of the coil's wire over the battery's positive (+) terminal so your finger does not touch the bare wire, as shown in Figure 2.

Place some small paper clips or screws near the screw.

Now when you press the loose wire on the battery's positive (+) terminal, the paper clips will cling to the steel screw since it becomes an electromagnet. See Figure 3.

Note: Do not keep this connected for more than a few seconds at a time.

FIGURE 1

FIGURE 2

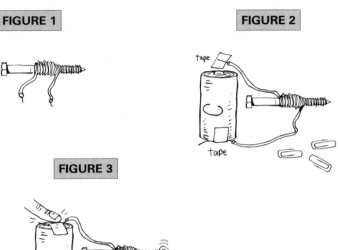

FIGURE 3

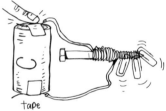

Sneaky Motor Fun

The electric motor is among the greatest inventions in modern history. Motors convert electrical energy into mechanical movement using the interaction of current (electron flow) and a magnetic field.

How exactly do motors spin? If you have played with magnets, you know that pushing the north end of one magnet towards the north end of another magnet causes the second magnet to move. An electric motor also does this using an electromagnetic force that is switched on and off to keep the motor's axle, or *shaft*, spinning around.

An electric motor is a cylinder packed with magnets around its edge. In the middle, there's a core made of iron wire wrapped around many times. When electricity flows into the iron core, it creates magnetism. The magnetism created in the core repels the magnets in the outer cylinder and makes the core of the motor spin.

The following projects illustrate how to make simple demonstration motors using everyday things. You'll also learn how to round up motors from discarded toys and devices and how to go further with other creative homemade motor designs.

Fun Fact: A motor performs the opposite function of a generator. You can prove it with the next project.

Make a Toy
Motor Generator

What's Needed
- Small toy car with motor
- LED
- Tape

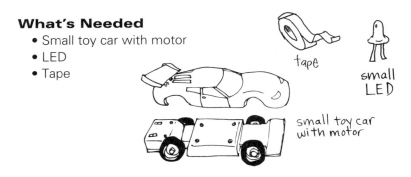

tape

small
LED

small toy car
with motor

What to Do
Remove the body from a small toy car's chassis as shown in
Figure 1.

Locate the electric motor and its two electrical terminals,
which are usually accessible on its case as shown in Figure 2.

Carefully tape the ends of the LED to the motor terminals
as shown in Figure 3. Now, spin the car wheel and you'll see
the LED should briefly light (if not, reverse the LED wires on the
motor and try it again).

If you still cannot see the LED light, test this procedure in a
dark room.

When you spin the car wheels, the motor's rotor (or rotating
part) moves a coil of wire near magnets, which generates an
electrical current. By spinning the wheels, you create enough
electrical power to light the LED.

In normal use, electrical power to the motor will cause the motor's wire coils (or electromagnets) to repel from the magnets and spin the shaft that is the axle for the car wheels.

Note: If you have a separate motor that is not mounted in a car (e.g. from an old electric toothbrush), you can spin the gear shaft rapidly with your fingers to light the LED.

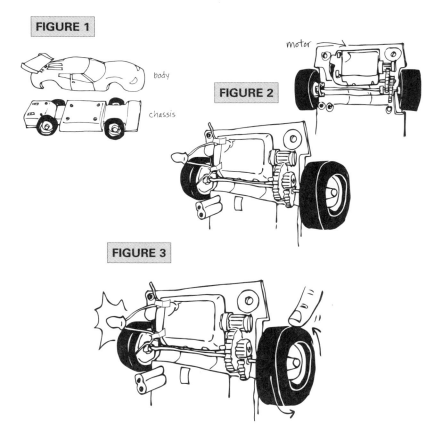

FIGURE 1

body

chassis

FIGURE 2

motor

FIGURE 3

Make a Sneaky Chip Can Motor

Let's see how you can make a simple motor with everyday things to demonstrate how to harness the power of electromagnetism.

METHOD 1

What's Needed

- Chip or peanut can with tin bottom
- Scouring pad
- 4 strong disc magnets
- 4 jumbo paperclips
- Cardboard food box
- 6 feet of stranded copper wire
- Battery holder that holds 4 C or D batteries
- 4 C or D batteries
- Bottle caps
- Tape

6 feet of stranded copper wire

battery holder and 4 "C" or "D" cell batteries

4 strong magnets

tape

4 paper clips

Food cardboard box

SUGARY Popthings

scouring pad

bottle caps

chip can

What to Do

First, scrape the chip can's tin bottom with a scouring pad to remove the protective coating.

Next, bend two paper clips into the dimensions shown in Figure 1.

Now, tape the two paper clips and tape them securely on the cardboard five inches apart.

Bend two more paper clips to act as axles for the chip can's rotor, as shown in Figure 3.

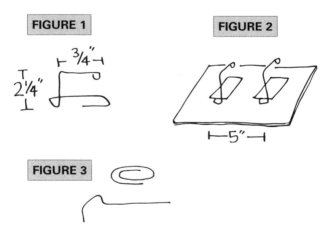

FIGURE 1

3/4"

2¼"

FIGURE 2

5"

FIGURE 3

Tape the two paper clip axles securely to the ends of the chip can and its plastic lid as shown in Figure 4.

Carefully tape four strong magnets to the center of the chip can spaced equally apart from one another, being sure the magnets are all facing in the same direction. First, mount one magnet, and then bring the next magnet close to it to see if it is attracted or repelled from it. Test all four magnets for the proper direction before taping them to the chip can. See Figure 5.

Next, place strips of tape aligned with where the magnets are mounted on the tin bottom of the can and slip the chip can, or rotor, in the paper clip stand as shown in Figure 6.

Wrap 30 turns of wire into a coil around a thick pen or magic marker, leave six inches free at each end and remove the pen. Tape the coil securely so it does not separate. Then, tape it on the cardboard between the paper clips, leaving the two ends free. See Figure 7.

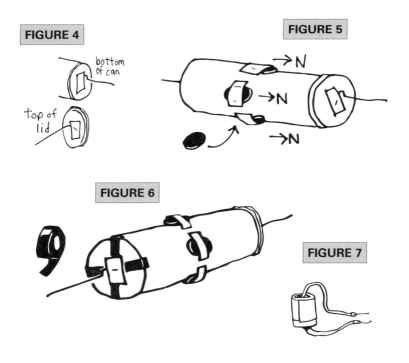

FIGURE 4

FIGURE 5

FIGURE 6

FIGURE 7

Connect one end of the wire coil to the battery pack and connect the other coil wire to the paper clip stand near the can's tin bottom. Then, connect a six-inch length of wire to the other battery pack terminal. This will act as the motor's contact wire, as shown in Figure 8.

Connect the batteries into the battery holder. Spin the can and lightly touch the contact wire end to the tin section. Experiment until you find the exact location where the chip can rotor continues to spin. Try spinning the can in the opposite direction. See Figure 9.

FIGURE 8

contact wire

FIGURE 9

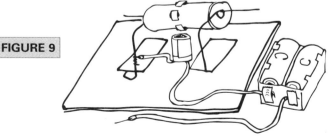

FIGURE 10

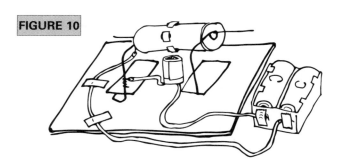

Once you find an area that allows the can to keep spinning on its own, tape the contact wire in place and adjust it so it lightly presses against the tin, as seen in Figure 10.

Once the best contact location is found, you can tape the contact wire in place so it lightly rests against the can, as shown in Figure 11.

FIGURE 11

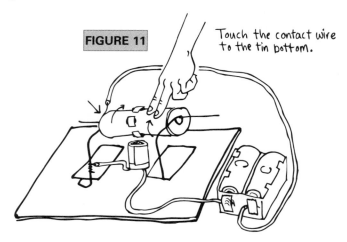

Touch the contact wire to the tin bottom.

How does this work? When the wire contacts the tin, it completes the electrical circuit and creates an electromagnetic field in the coil, which repels the magnets on the can away causing it to spin. When the can spins, the contact wire touches the tape disconnecting the magnetic field, allowing the can to keep spinning. Without the tape, the can would stop.

METHOD 2

What's Needed
- Aluminum foil
- Cardboard tube
- 4 strong disc magnets
- 4 jumbo paper clips
- Cardboard food box
- 6 feet of stranded copper wire
- Battery holder that holds 4 C or D batteries
- 4 C or D batteries
- Bottle caps
- Tape

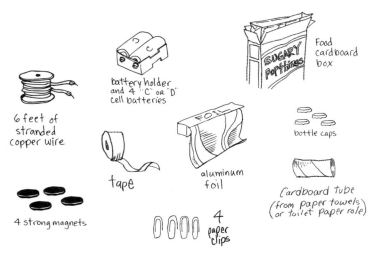

6 feet of stranded copper wire

battery holder and 4 "C" or "D" cell batteries

Food cardboard box

bottle caps

tape

aluminum foil

Cardboard tube (from paper towels or toilet paper role)

4 strong magnets

4 paper clips

What to Do

You can make another version of this motor using a plain cardboard tube instead of a can with a tin bottom by adding a strip of aluminum foil to one end of the tube.

First, construct the paper clip stands, wire coil, and battery pack as shown in the previous motor project.

You can use a paper towel tube or a toilet paper roll—or make a tube with cardboard—and tape the magnets as shown.

Trace circles on the cardboard, using the end of the cardboard tube as a guide. These will act as end caps for the rotor. See Figure 1.

Draw four "tabs" on the circles to allow you to fold them over the end of the tube and affix them with tape as shown in Figure 2.

Bend the paper clips into the shape shown in Figure 3.

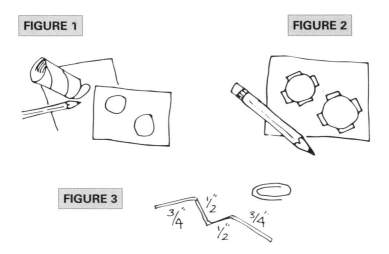

FIGURE 1

FIGURE 2

FIGURE 3

Tape four strong disc magnets, spaced equally apart, at the center of the tube. See Figure 4.

Cut out and place the cardboard end caps on the ends of the tube, fold down the tabs, and tape them to the tube. See Figure 5.

Tape the paper clip axles on the end caps as shown in Figure 6.

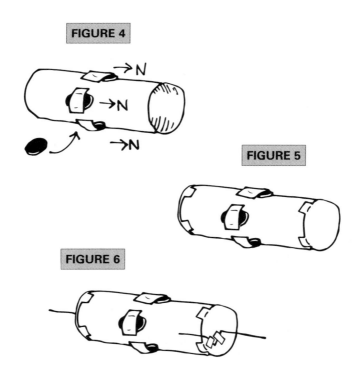

FIGURE 4

FIGURE 5

FIGURE 6

Tape the aluminum foil strip on one end of the tube and ensure that it makes good contact with the paper clip axle as shown in Figure 7.

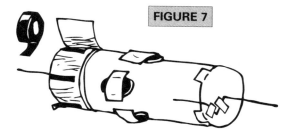

FIGURE 7

Cut thin strips of tape aligned with each magnet as shown in Figure 8. Mount the tube on the paper clip axles.

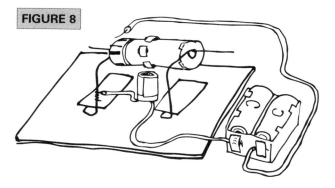

FIGURE 8

As with the previous motor project, spin the tube and touch the free wire on the aluminum foil in various spots until you locate a spot where the motor keeps turning on its own, and then secure the wire as shown in Figures 9 and 10.

FIGURE 9

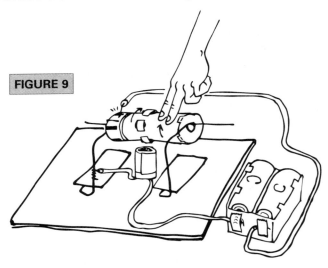

FIGURE 10

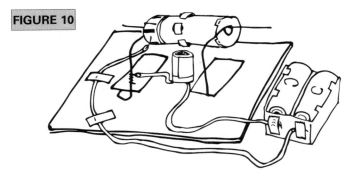

How does this work? When the wire contacts the aluminum foil, it creates an electromagnet in the wire coil and repels the magnets on the cardboard tube away from it. When the can spins, the wire touches the tape and disconnects the magnetic field, allowing the can to keep spinning. Without the tape, the can would just stop.

Sneaky Motor Troubleshooting Tips

- Ensure your batteries are fresh.
- Ensure the wire coil is **directly beneath** the magnets and **as close as possible** to them without touching.
- Reverse the wires attached to the battery holder.
- Use stronger magnets; neodymium disc magnets are recommended for making homemade motors.
- Ensure the rotor does not slide (left or right) on the stands too far away from the coil beneath. You can apply a small width of tape on the paper clip axle or slip a short piece of wire insulation over it to prevent sliding.
- If the rotor does not spin with the contact wire, adjust the area where it makes contact.
- Try spinning the rotor in the reverse direction and search for the best loose wire contact spot.
- Touch the loose contact wire to the paper clip stand to see if the rotor moves; if not, tighten the wire connecting from the coil to the stand.
- If the motor spins for a while and stops, carefully clean off the tin can bottom or the tube rotor's aluminum foil. Eventually the aluminum foil will have to be replaced with a new strip.

Going Further with Sneaky Motor-Making

You can continue improving your motor designs with these creative operation and design ideas:

- See if you can spin the rotor, and keep it spinning, by motioning with a magnet in your hand, repelling the rotor magnets.
- Experiment with the position of the tape strips—use thicker widths or move them a small distance left or right away from the magnets.
- Add additional magnets and tape strips.
- Make a sturdier platform and mounts for the rotor.
- Add a disc to act as a flywheel on the end of the axle to store energy.

Sneaky Motor Design Ideas and Additions

- Add a second coil wired in series with the original one and position them opposite the motor's rotor.

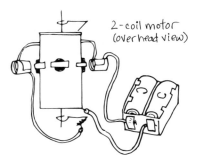

2-coil motor (overhead view)

- Design and make a horizontal platter motor.

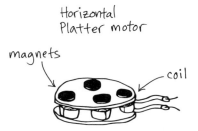

Horizontal Platter motor

magnets

coil

• Design and make a solenoid motor using a cranking arm.

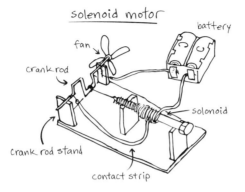

solenoid motor

battery

fan

crank rod

solenoid

crank rod stand

contact strip

• Design and make a motor with the coils on the rotor and the magnet mounted on the base.

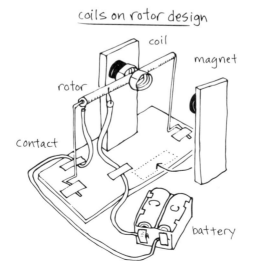

coils on rotor design

coil

magnet

rotor

contact

battery

- Attach an automaton device (a self-operating toy normally powered with a hand crank) at the end of a modified motor axle.

crank handle

Automaton

- Disassemble small motors found in discarded toys and electric toothbrushes to study their designs and salvage strong magnets as shown in "How to Disassemble a Small Toy Motor."

How to Disassemble a Small Toy Motor

Figure 1 shows a typical small toy motor. Its diameter is about the same as a dime. Its end cap is plastic and has two contacts on it for connecting power.

If it has a gear or weight on the axle, pull it off with pliers. A weight is usually mounted off-center to make the motor vibrate. See Figure 2.

To remove the plastic end cap, pry up the case metal tabs as shown in Figure 3.

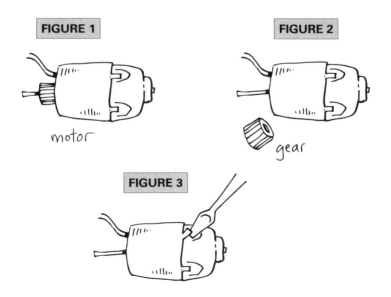

FIGURE 1

motor

FIGURE 2

gear

FIGURE 3

Figure 4 shows the end cap removed from the motor. The brushes are mounted on the inside of the end cap and they are connected to the contacts on the other side of the end cap. The brushes connect power to the motor's rotor.

You can now slide the rotor out of the case. On the rotor are usually three or more coils called *armatures*. See Figure 5.

Next, you can remove the magnets on the inside of the case. They may slide out or have a "V" shaped spring clip holding them in that you can remove with pliers and a small screwdriver. See Figure 6.

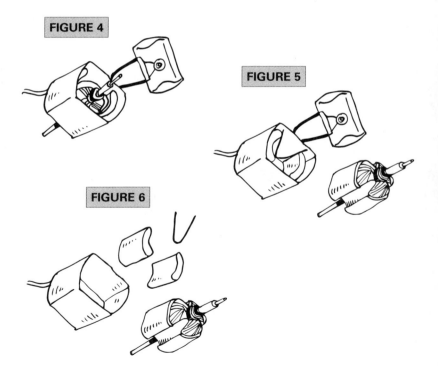

FIGURE 4

FIGURE 5

FIGURE 6

Note: If the magnets were held in the case with a spring clip, you may have difficulty replacing the spring clip.

Fun Trick: First, slide the magnets into the case. Then, slide the rotor inside. Now it should be easier to compress and slide the spring clip inside. To be sure, slide out the rotor and check if the magnets and spring clip are secure. Then slide the rotor and end cap in place and push down the metal tabs on the case. See Figure 7.

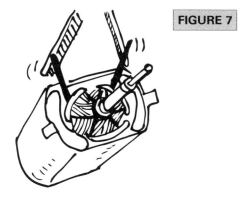

FIGURE 7

Sneaky Motor Bonus Project:
Convert a Toothbrush into a Vibrabot

Inexpensive electric toothbrushes, available at discount and "dollar" stores, include the parts you need to make a vibrabot toy vehicle that vibrates across a surface. (You don't have to have an electric toothbrush for this project—it can be completed using virtually any small toy motor, wire, battery, and cardboard.)

What's Needed

- Electric toothbrush
- AA battery
- 2 twist ties or 6 inches of copper wire
- Tape
- Pliers

twist ties

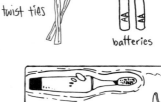

batteries

tape

pliers

What to Do

First, separate the product packaging; remove the clear plastic cover from the cardboard as shown in Figure 1.

Cut a small piece of cardboard approximately two inches by three inches. Then cut two pieces of raised molded plastic to act as the legs of the Vibrabot. See Figure 2.

Tape the plastic legs to the bottom of the cardboard, as shown in Figure 3.

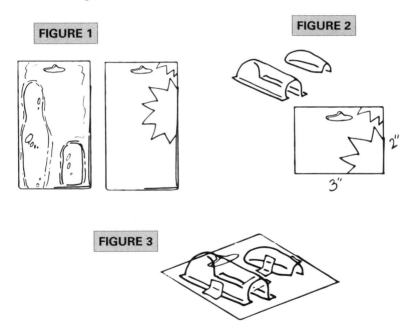

FIGURE 1

FIGURE 2

2"

3"

FIGURE 3

Figure 4 shows the motor and battery holder that you can pull out of the bottom of the toothbrush with pliers.

The motor and battery holder will separate by holding them in your hands and twisting. See Figure 5.

Strip the ends of the twist ties (or wire) to expose the bare wire as shown in Figure 6.

Tape the AA battery on top of the motor with the positive (+) terminal facing in the direction of the front of the motor. See Figure 7.

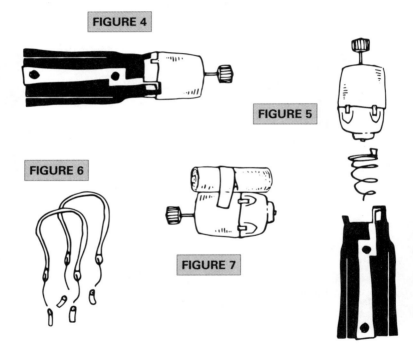

FIGURE 4

FIGURE 5

FIGURE 6

FIGURE 7

Next, connect the two twist ties to the motor contacts as shown in Figure 8.

Tape the motor and battery to the top of the cardboard. See Figure 9.

FIGURE 8

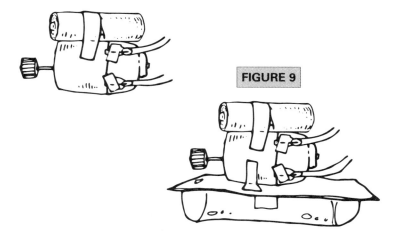

FIGURE 9

As shown in Figure 10, tape one twist tie to the battery's negative (-) terminal and tape the remaining twisttie to the top of the battery near its positive (+) terminal.

When you bend and press the twist-tie onto the battery's positive (+) terminal, the Vibrabot will scurry about on a flat surface. To stop it, bend the wire back on top of the battery.

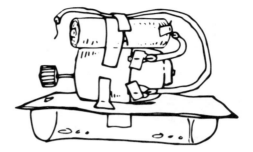

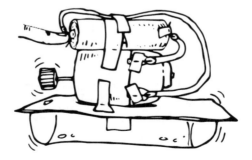

Going Further:
Sneaky Motor Design Ideas and Additions

Gather discarded motors, and add a battery and switch to make: toy cars, vibrabot toys, water pump, vacuum cleaner, miniature drill, bubble blower, motorized drink mixer, pulley system (for slow speed carousels and automaton designs as shown on the next pages.)

Motorized bottle cap car

Vacuum cleaner

Water pump

Motorized drill

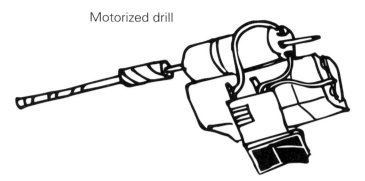

Bubble blower

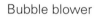

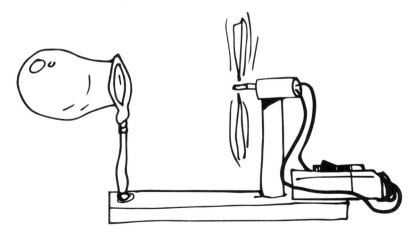

Motorized drink
mixer

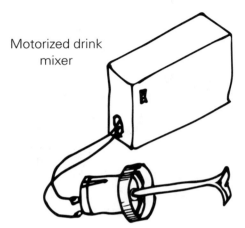

Pulley system (for slow-speed automaton projects)

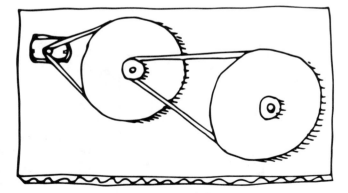

Sample automatons

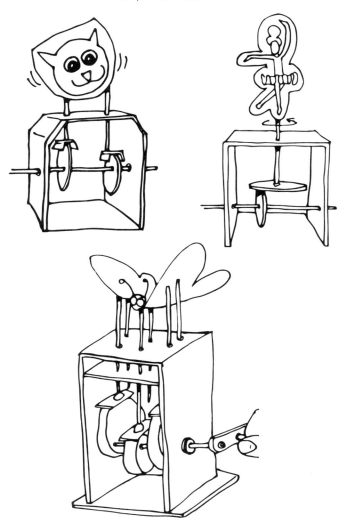

RESOURCES

You can always do more with what's around you. Indeed, one day it may be necessary for you to improvise to survive. Look around you and think about the many ways you can put other items to alternative use. What can you make with a radio control car, walkie-talkie, windup toys, retractable badge holders, toy dart guns, toy robots, or rubber band-powered items? Plenty.

Think of the many ways a simple item like a flashlight can be used. Ponder its every part. Here are just a few sneaky functions:

The case can be used to collect water, to block a window from opening, as a safe for small valuables, as a fire extinguisher container, and as a weapon.

The reflector can be used to start a fire or to signal for help.

The bulb's shell can cut small items.

The batteries can be used to start a fire or to magnetize a piece of metal.

The lens cover can be used with water as a magnifier or to start a fire.

The bulb filament can be used to start a fire.

What other parts of the flashlight have not been mentioned? What can they be used for? Your way of looking at things is your greatest resource. To borrow a business phrase, "Think outside the box." The projects in this book should have expanded your thoughts about what you can do with everyday things.

Note: Be sure to visit **Sneakyuses.com** for extra project information, web links, and resourceful contest information and to post your imaginative discoveries.

USEFUL WEBSITES

Frugal and Thrift Sites

choose2reuse.co.uk
make-stuff.com
Recycle.net
thefrugalshopper.com
wackyuses.com

Pocket Tools and Gadgets

advanced-intelligence.com
colibri.com
leatherman.com
scientificsonline.com
spyderco.com
swissarmy.com
swisstechtools.com
topeak.com

Survival

backwoodsmanmag.com
basegear.com
beprepared.com
campmor.com
equipped.com
fieldandstream.com
ruhooked.com
Survival.com
survival-center.com
Survivaliq.com

Home Security Sites

ncpc.org

X10.com®

Science and Technology

HomeAutomationMag.com

howstuffworks.com

lifewire.com

midnightscience.net

scienceproject.com

Scientificsonline.com

scitoys.com

gamestop.com/collection/thinkgeek

Other Websites of Interest

Almanac.com

Doityourself.com

Instructables.com

Makezine.com

Popsci.com

Popularmechanics.com

rube-goldberg.com

Smarthome.com

tbotech.com

Tipking.com

RECOMMENDED READING

Books

Ira Flatow, *They All Laughed . . . From Light Bulbs to Lasers: The Fascinating Stories Behind the Great Inventions That Have Changed Our Lives* (Harper Perennial)

Joey Green, *Clean It! Fix It! Eat It!: Easy Ways to Solve Everyday Problems with Brand-Name Products You've Already Got Around the House* (Prentice Hall)

———, *Clean Your Clothes with Cheez Wiz: And Hundreds of Offbeat Uses for Dozens More Brand-Name Products* (Prentice Hall)

———, *Joey Green's Encyclopedia of Offbeat Uses for Brand-Name Products* (Prentice Hall)

Lois H. Gresh and Robert Weinberg, *The Science of Superheroes* (John Wiley & Sons)

William Gurstelle, *Backyard Ballistics* (Chicago Review Press)

Garth Hattingh, *The Outdoor Survival Handbook* (New Holland Publishers)

Dave Hrynkiw and Mark W. Tilden, *JunkBots, Bugbots, and Bots on Wheels: Building Simple Robots With BEAM Technology* (McGraw-Hill Osborne Media)

Vicky Lansky, *Another Use for 101 Common Household Items* (Book Peddlers)

———. *Baking Soda: Over 500 Fabulous, Fun, and Frugal Uses* (Book Peddlers)

———, *Don't Throw That Out: A Pennywise Parent's Guide* (Book Peddlers)

———, *Transparent Tape: Over 350 Super, Simple, and Surprising Uses* (Book Peddlers)

Joel Levy, *Really Useful: The Origins of Everyday Things* (Firefly Books)

Hugh McManners, *The Complete Wilderness Training Book* (Dorling Kindersley)

Forrest M. Mims III, *Circuits and Projects* (Radio Shack)

———, *Science and Communications Circuits and Projects* (Radio Shack)

Steven W. Moje, *Paper Clip Science* (Sterling Publishing Co.)

Bob Newman, *Wilderness Wayfinding: How to Survive in the Wilderness as You Travel* (Paladin Press)

Tim Nyberg and Jim Berg, *The Duct Tape Book* (Workman Publishing Company, 1994)

———, *Duct Tape Book Two: Real Stories* (1995)

———, *The Ultimate Duct Tape Book* (1998)

Larry Dean Olsen, *Outdoor Survival Skills* (Chicago Review Press)

Robert Young Pelton, *Come Back Alive* (Doubleday)

Joshua Piven and David Borgenicht, *The Worst Case Scenario Survival* (Chronicle Books)

———, *The Worst Case Scenario Travel* (Chronicle Books)

Royston M. Roberts, *Serendipity* (John Wiley & Sons)

US Air Force Search and Rescue Handbook (Lyons Press)

US Army Survival Handbook (Lyons Press)

Jim Wilkinson and Neil A. Downie, *Vacuum Bazookas, Electric Rainbow Jelly, and 27 Other Saturday Science Projects* (Princeton University Press)

John Wiseman, *The SAS Survival Handbook* (Harvill Books)

Joey Green Books, *Last-Minute Survival Secrets: 128 Ingenious Tips to Endure the Coming Apocalypse and Other Minor Inconveniences*

———. *Potato, Dizzy Dice, and More Wacky, Weird Experiments from the Mad Scientist*

———. *The Ultimate Mad Scientist Handbook*

Sean Connelly Books, *The Book of Totally Irresponsible Science: 64 Daring Experiments for Young Scientists*
——. *The Book of Ingeniously Daring Chemistry: 24 Experiments for Young Scientists*
——. *The Book of Terrifyingly Awesome Technology: 27 Experiments for Young Scientists*

Magazines

Backpacker
E: The Environmental Magazine
Mother Earth News
Nuts and Volts
Outdoor Life
Outside
Popular Mechanics
Popular Science